Being and the Screen

Design Thinking, Design Theory

Ken Friedman and Erik Stolterman, editors

Design Things, A. Telier (Thomas Binder, Pelle Ehn, Giorgio De Michelis, Giulio Jacucci, Per Linde, and Ina Wagner), 2011

China's Design Revolution, Lorraine Justice, 2012

Adversarial Design, Carl DiSalvo, 2012

The Aesthetics of Imagination in Design, Mads Nygaard Folkmann, 2013

Linkography: Unfolding the Design Process, Gabriela Goldschmidt, 2014

Situated Design Methods, edited by Jesper Simonsen, Connie Svabo, Sara Malou Strandvad, Kristine Samson, Morten Hertzum, and Ole Erik Hansen, 2014

Taking [A]part: The Politics and Aesthetics of Participation in Experience-Centered Design, John McCarthy and Peter Wright, 2015

Design, When Everybody Designs: An Introduction to Design for Social Innovation, Ezio Manzini, 2015

Frame Innovation: Creating New Thinking by Design, Kees Dorst, 2015

Designing Publics, Christopher A. Le Dantec, 2016

Overcrowded: Designing Meaningful Products in a World Awash with Ideas, Roberto Verganti, 2016

FireSigns: A Semiotic Theory for Graphic Design, Steven Skaggs, 2017

Making Design Theory, Johan Redström, 2017

Critical Fabulations: Reworking the Methods and Margins of Design, Daniela Rosner, 2018

Designing with the Body: Somaesthetic Interaction Design, Kristina Höök, 2018

Discursive Design: Critical, Speculative, and Alternative Things, Bruce M. Tharp and Stephanie M. Tharp, 2018

Pretense Design: Surface over Substance, Per Mollerup, 2019

Being and the Screen: How the Digital Changes Perception, Stéphane Vial, 2019

Being and the Screen

How the Digital Changes Perception

Published in one volume with *A Short Treatise on Design*

Stéphane Vial

translated by Patsy Baudoin

The MIT Press
Cambridge, Massachusetts
London, England

© 2019 Massachusetts Institute of Technology

Originally published in France by Presses Universitaires de France/Humensis:
L'Être et l'écran, 2nd ed. © 2017, Presses Universitaires de France/Humensis
Court traité du design, 2nd ed. © 2014, Presses Universitaires de France/Humensis

This book is published with the support of the University of Nîmes, France.

All rights reserved. No part of this book may be reproduced in any form by any electronic or mechanical means (including photocopying, recording, or information storage and retrieval) without permission in writing from the publisher.

This book was set in ITC Stone Serif Std and ITC Stone Sans Std by Toppan Best-set Premedia Limited. Printed and bound in the United States of America.

Library of Congress Cataloging-in-Publication Data

Names: Vial, Stéphane, author. | Baudoin, Patsy, translator. | Folkmann, Mads Nygaard, 1972– writer of supplementary textual content. | Lévy, Pierre, writer of supplementary textual content.
Title: Being and the screen : how the digital changes perception : published in one volume with A short treatise on design / Stéphane Vial ; translated by Patsy Baudoin.
Other titles: L'être et l'écran. English
Description: Cambridge, MA : MIT Press, [2019] | Series: Design thinking, design theory | Includes bibliographical references and index.
Identifiers: LCCN 2019005611 | ISBN 9780262043168 (hardcover : alk. paper)
Subjects: LCSH: Technology—Philosophy. | Design—Philosophy. | Digital electronics—Psychological aspects. | Perception (Philosophy)
Classification: LCC T14 .V53413 2019 | DDC 601—dc23 LC record available at https://lccn.loc.gov/2019005611

10 9 8 7 6 5 4 3 2 1

For Laëtitia, because I met her on-screen

The computer is an enigma. Not in its making or its usage, but because man appears incapable of foreseeing anything about the computer's influence on society and humanity.
—Jacques Ellul, *The Technological System* (1977)

For me, the computer is the most remarkable tool we invented. It's the equivalent of the bicycle for the mind.
—Steve Jobs in Julian Krainin and Michael R. Lawrence, *Memory and Imagination: New Pathways to the Library of Congress* (1990)

Computers don't just do things for us, they do something to us.
—Sherry Turkle, *Life on the Screen: Identity in the Age of the Internet* (1995)

Contents

Series Foreword xi
Preface to the US Edition xvii

I Being and the Screen: How the Digital Changes Perception 1

Foreword to the First Edition 3
 Critic and Visionary: The Double Gaze of the Humanities
 Pierre Lévy

Introduction: What Is the Digital Revolution Revolutionizing? 7

1 Technology as a System 11

2 The Digital Technological System 27

3 The Technological Structures of Perception 41

4 The Life and Death of the Virtual 65

5 Digital Ontophany 81

6 The (Digital) Design of Experience 111

Conclusion: On the Radical Aura of Things 125

Supplement 1: Otherphany and Otherness 131

Supplement 2: Ontophanic Feeling 137

Supplement 3: Against Digital Dualism: A Phenomenological Critique of Judgment 141

II   A Short Treatise on Design   147

Foreword to the First Edition   149
Raising the Question of Design
Mads Nygaard Folkmann

1   The Paradox of Design: Wherein We Show That Design Thinks but Does Not Reflect upon Itself   153

2   The Disorder of Speech: Wherein We Deconstruct and Rebuild the Word *Design*   155

3   Design, Crime, and Marketing: Wherein We Talk about the Very Horrific Alliance between Design and Capital   163

4   Beyond Capital: Wherein We State the Moral Law of the Designer   169

5   The Design Effect: Wherein We Reduce the Quiddity of Design to Three Criteria   173

6   Drafting a Project: Wherein We Show That the Designer Is Not an Artist   179

7   Design as "A Thing That Thinks": Wherein We Defend the Concept of "Design Thinking"   183

8   Toward Digital Design: Wherein We Look into the Consequences of the Interactive Revolution   187

Postscript: The Design System: Wherein the Author Orders His Principles in Geometric Style   197
Notes   201
Bibliography   231
Index   241

Series Foreword

As professions go, design is relatively young. The practice of design predates professions. In fact, the practice of design—making things to serve a useful goal, making tools—predates the human race. Making tools is one of the attributes that made us human in the first place.

Design, in the most generic sense of the word, began over 2.5 million years ago when *Homo habilis* manufactured the first tools. Human beings were designing well before we began to walk upright. Four hundred thousand years ago, we began to manufacture spears. By forty thousand years ago, we had moved up to specialized tools.

Urban design and architecture came along ten thousand years ago in Mesopotamia. Interior architecture and furniture design probably emerged with them. It was another five thousand years before graphic design and typography got their start in Sumeria with the development of cuneiform. After that, things picked up speed.

All goods and services are designed. The urge to design—to consider a situation, imagine a better situation, and act to create that improved situation—goes back to our prehuman ancestors. Making tools helped us to become what we are—design helped to make us human.

Today, the word "design" means many things. The common factor linking them is service, and designers are engaged in a service profession in which the results of their work meet human needs.

Design is first of all a process. The word "design" entered the English language in the 1500s as a verb, with the first written citation of the verb dated to the year 1548. *Merriam-Webster's Collegiate Dictionary* defines the verb "design" as "to conceive and plan out in the mind; to have as a specific purpose; to devise for a specific function or end." Related to these is the act of drawing, with an emphasis on the nature of the drawing as a plan or map, as well as "to draw plans for; to create, fashion, execute or construct according to plan."

Half a century later, the word began to be used as a noun, with the first cited use of the noun "design" occurring in 1588. *Merriam-Webster's* defines the noun as "a particular purpose held in view by an individual or group; deliberate, purposive planning; a mental project or scheme in which means to an end are laid down." Here, too, purpose and planning toward desired outcomes are central. Among these are "a preliminary sketch or outline showing the main features of something to be executed; an underlying scheme that governs functioning, developing or unfolding; a plan or protocol for carrying out or accomplishing something; the arrangement of elements or details in a product or work of art." Today, we design large, complex processes, systems, and services, and we design organizations and structures to produce them. Design has changed considerably since our remote ancestors made the first stone tools.

At a highly abstract level, Herbert Simon's definition covers nearly all imaginable instances of design. To design, Simon writes, is to "[devise] courses of action aimed at changing existing situations into preferred ones" (Simon, *The Sciences of the Artificial*, 2nd ed., MIT Press, 1982, p. 129). Design, properly defined, is the entire process across the full range of domains required for any given outcome.

But the design process is always more than a general, abstract way of working. Design takes concrete form in the work of the service professions that meet human needs, a broad range of making and planning disciplines. These include industrial design, graphic design, textile design, furniture design, information design, process design, product design, interaction design, transportation design, educational design, systems design, urban design, design leadership, and design management, as well as architecture, engineering, information technology, and computer science.

These fields focus on different subjects and objects. They have distinct traditions, methods, and vocabularies, used and put into practice by distinct and often dissimilar professional groups. Although the traditions dividing these groups are distinct, common boundaries sometimes form a border. Where this happens, they serve as meeting points where common concerns build bridges. Today, ten challenges uniting the design professions form such a set of common concerns.

Three performance challenges, four substantive challenges, and three contextual challenges bind the design disciplines and professions together as a common field. The performance challenges arise because all design professions:

1. Act on the physical world;
2. Address human needs; and
3. Generate the built environment.

In the past, these common attributes were not sufficient to transcend the boundaries of tradition. Today, objective changes in the larger world give rise to four substantive challenges that are driving convergence in design practice and research. These substantive challenges are:

1. Increasingly ambiguous boundaries between artifacts, structure, and process;
2. Increasingly large-scale social, economic, and industrial frames;
3. An increasingly complex environment of needs, requirements, and constraints; and
4. Information content that often exceeds the value of physical substance.

These challenges require new frameworks of theory and research to address contemporary problem areas while solving specific cases and problems. In professional design practice, we often find that solving design problems requires interdisciplinary teams with a transdisciplinary focus. Fifty years ago, a sole practitioner and an assistant or two might have solved most design problems; today, we need groups of people with skills across several disciplines, and the additional skills that enable professionals to work with, listen to, and learn from each other as they solve problems.

Three contextual challenges define the nature of many design problems today. While many design problems function at a simpler level, these issues affect many of the major design problems that challenge us, and these challenges also affect simple design problems linked to complex social, mechanical, or technical systems. These issues are:

1. A complex environment in which many projects or products cross the boundaries of several organizations, stakeholder, producer, and user groups;

2. Projects or products that must meet the expectations of many organizations, stakeholders, producers, and users; and
3. Demands at every level of production, distribution, reception, and control.

These ten challenges require a qualitatively different approach to professional design practice than was the case in earlier times. Past environments were simpler. They made simpler demands. Individual experience and personal development were sufficient for depth and substance in professional practice. While experience and development are still necessary, they are no longer sufficient. Most of today's design challenges require analytic and synthetic planning skills that cannot be developed through practice alone.

Professional design practice today involves advanced knowledge. This knowledge is not solely a higher level of professional practice. It is also a qualitatively different form of professional practice that emerges in response to the demands of the information society and the knowledge economy to which it gives rise.

In a recent essay ("Why Design Education Must Change," *Core77*, November 26, 2010), Donald Norman challenges the premises and practices of the design profession. In the past, designers operated on the belief that talent and a willingness to jump into problems with both feet gives them an edge in solving problems. Norman writes:

> In the early days of industrial design, the work was primarily focused upon physical products. Today, however, designers work on organizational structure and social problems, on interaction, service, and experience design. Many problems involve complex social and political issues. As a result, designers have become applied behavioral scientists, but they are woefully undereducated for the task. Designers often fail to understand the complexity of the issues and the depth of knowledge already known. They claim that fresh eyes can produce novel solutions, but then they wonder why these solutions are seldom implemented, or if implemented, why they fail. Fresh eyes can indeed produce insightful results, but the eyes must also be educated and knowledgeable. Designers often lack the requisite understanding. Design schools do not train students about these complex issues, about the interlocking complexities of human and social behavior, about the behavioral sciences, technology, and business. There is little or no training in science, the scientific method, and experimental design.

This is not industrial design in the sense of designing products, but industry-related design, design as thought and action for solving problems and imagining new futures. This new MIT Press series of books emphasizes strategic design to create value through innovative products and services, and it emphasizes design as service through rigorous creativity, critical inquiry, and an ethics of respectful design. This rests on a sense of understanding, empathy, and appreciation for people, for nature, and for the world we shape through design. Our goal as editors is to develop a series of vital conversations that help designers and researchers to serve business, industry, and the public sector for positive social and economic outcomes.

We will present books that bring a new sense of inquiry to the design, helping to shape a more reflective and stable design discipline able to support a stronger profession grounded in empirical research, generative concepts, and the solid theory that gives rise to what W. Edwards Deming described as profound knowledge (Deming, *The New Economics for Industry, Government, Education*, MIT, Center for Advanced Engineering Study, 1993). For Deming, a physicist, engineer, and designer, profound knowledge comprised systems thinking and the understanding of processes embedded in systems; an understanding of variation and the tools we need to understand variation; a theory of knowledge; and a foundation in human psychology. This is the beginning of "deep design"—the union of deep practice with robust intellectual inquiry.

A series on design thinking and theory faces the same challenges that we face as a profession. On one level, design is a general human process that we use to understand and to shape our world. Nevertheless, we cannot address this process or the world in its general, abstract form. Rather, we meet the challenges of design in specific challenges, addressing problems or ideas in a situated context. The challenges we face as designers today are as diverse as the problems clients bring us. We are involved in design for economic anchors, economic continuity, and economic growth. We design for urban needs and rural needs, for social development and creative communities. We are involved with environmental sustainability and economic policy, agriculture competitive crafts for export, competitive products and brands for micro-enterprises, developing new products for bottom-of-pyramid markets and redeveloping old products for mature or wealthy markets. Within the framework of design, we are also challenged to design for extreme situations, for biotech, nanotech, and new materials, and design for social business, as well as conceptual challenges for worlds that do not yet exist such as the world beyond the Kurzweil singularity—and for new visions of the world that does exist.

The Design Thinking, Design Theory series from the MIT Press will explore these issues and more—meeting them, examining them, and helping designers to address them.

Join us in this journey.

Ken Friedman and Erik Stolterman
Editors, Design Thinking, Design Theory Series

Preface to the US Edition

Two books undertaking the same challenge make up this book: to build a phenomenology of technology and design in order to cast light on our experiences relating to artifacts, especially digital artifacts, and to grasp the philosophical meaning of the act of design.

*

First published in 2013 with the Presses Universitaires de France, *Being and the Screen* stems from my doctoral thesis in philosophy, completed at the Sorbonne at Paris Descartes University. Reissued and extended in 2017, this book has garnered many invitations for me in the most diverse circles, including one from the president of the French Republic, François Hollande, to a lunch about the digital at the Elysée Palace in Paris. There are few books on the philosophy of technology that have had such a path.

I see this as a sign that *Being and the Screen* answers to a profound intellectual need: to offer philosophical meaning to the digital revolution on a level other than the most common one of analyzing technology through economic and political lenses. *Being and the Screen* is not about digital capitalism and carefully tries to avoid being about it. It has no interest in the economico-political dimensions of the digital, that is, of the digital in its relation to the individual and society. Rather, it inscribes its effort in the epistemological analysis of technologies, and it deals with the digital in relation to subjects and objects, that is, in its anthropological-phenomenological dimension, considered as the substratum or the basis of all other dimensions.

Being and the Screen tries to introduce a new approach to the "historical phenomenology of technologies," which I also call the "phenomenological archaeology of technologies." This is a new branch of the phenomenology of technologies that is parallel and complementary to that of postphenomenology. It is based on ontophany theory, according to which the process of appearance of being is constantly technologically conditioned. Thus, technologies are not only tools; they are structures of perception. They are apparatuses in which appearance makes its appearance. In recent years, this theory has been supplemented by other publications and has been extended by other researchers in France, for example, by Samira Ibnelkaïd in her remarkable, empirical work on multimodal screen interactions.[1]

This American edition of *Being and the Screen* is a translation of the second, revised and enriched edition published in 2017. It has been subject to some clarifications and adaptations for the US and English public, but it remains intact in its essence and is relatively timeless. The theory of ontophany is indeed a lasting philosophical solution to many current debates on the digital, such as the fear of screens.[2]

The fear of screens is in a sense a normal and inevitable phenomenological anxiety owing to the slow acculturation of our perceptual structures to the new digital ontophany of the world. The phenomenological trauma of the digital is not so easy to absorb. A background of "IRL fetishism" persists in all of us.[3] It will take time, years and decades, to fully integrate the digital into the lens of our perception. Working on oneself, meditating on one's own ontophanic transformation, however, can speed things up. By this, I mean a kind of contemporary "spiritual exercise,"[4] which I understand as a work of phenomenological self-analysis, whereby each of us can observe ourself and try to understand our own ontophanic process. When we succeed, we become sensitive to the technoperceptual aspect of presence, and we get in touch with ontophanic feeling, which alone allows us to get some necessary distance. A French researcher recently told me, "Your chapter on technology as ontophanic matrix is really changing my way of perceiving my environment. I cannot explain it yet, but that will come." That's all that I wish for you!

*

A Short Treatise on Design appeared in 2010, before *Being and the Screen*, with the Presses Universitaires de France, after I taught philosophy for five years at École Boulle, a well-known Parisian school of art and design. Revised in 2014 for the second edition, this book was conceived as an introduction to

Preface to the US Edition xix

the possible philosophical dimension of design. It has been translated into Swedish (2011), Korean (2012), and Chinese (2017). This is the book of a philosopher who discovers design in a design school, a philosopher among designers, introduced to the world of studios and the practice of projects, and who will then in turn design—interaction design—in his own studio in Paris.

I am no longer at École Boulle or professionally doing web design. I am now in academia, engaged in research about design, and my ideas on design have evolved. Yet they remain largely faithful to the intuitions in this book. Deliberately short, *A Short Treatise on Design* is above all an introduction to the discipline of design—not a history book, but a book built of philosophical problems. The philosophy of design is still an emerging discipline,[5] but this book's modest contribution lies in the theory of the effect of design. This theory proposes that we think about design not as producing *beings* but, rather, *events*; not *things that are* but *things that happen*; what I call *effects*—along three axes: ontophanic (effect of experience), callimorphic (effect of formal beauty), and socioplastic (societal effect). That is precisely where the link is between *A Short Treatise on Design* and *Being and the Screen*. Design is creative work on ontophany. Creative phenomenology is what produces neophanies, novel ways of appearing and structuring lived experience. Doing design is therefore to be philosophically responsible for existential experience.

*

I cannot finish this preface without warmly thanking the people without whom the publication of this book would not have been possible or who contributed in one way or another to making its publication possible. I am thinking of Erik Stolterman and Ken Friedman, Douglas Sery and Maria Vlachou, Neal Stimler and Lev Manovich, Pierre Lévy and Mads N. Folkmann. I also thank Patsy Baudoin for her outstanding work of translation. And finally, a personal thank you to Laëtitia for her patience and encouragement.

Stéphane Vial
Montreal, Canada
November 2018
www.stephane-vial.net

I Being and the Screen: How the Digital Changes Perception

Foreword to the First Edition
Critic and Visionary: The Double Gaze of the Humanities

Pierre Lévy

If we want to see a little more clearly, beyond the simple posture of consumers or users, how we integrate our thoughts and symbols in the algorithmic medium, if we want to understand the digital mutation going on, and if we want to give ourselves the means of weighing in on its progress, we have to keep both eyes wide open—our critical eye as well as our visionary eye.

When it comes to the critical eye, let's first learn to smile when facing junky slogans; marketers' click bait, competition for Klout scores,[1] and "open rebel" poses. For some, the internet may be a new religion. Why not? But for heaven's sake, let's not build new idols: the internet is not an actor or a source of information, not a universal solution or a model. (Evgeny Morozov explains all of this nicely in his latest book, *To Save Everything, Click Here*.)[2]

The internet is not an actor. The new algorithmic medium that is becoming increasingly more complex under our very eyes and fingers is certainly not a homogeneous actor, but rather the hypercomplex assemblage of multitudinous human and nonhuman actors of all kinds, an assemblage that is constantly and rapidly changing, a metamedium that houses and interweaves a great variety of media, each of which calls for a particular analysis in a particular sociohistorical context. The algorithmic medium does not make decisions and does not act independently.

Nor is it a source of information: only the people and institutions that express themselves there are actual sources. Many journalists maintain the confusion that stems from the fact that in traditional media with unilateral dissemination (organs of the press, radio, television), the channel merges with the transmitter. But in the new communications environment, many independent sources may use the same platform.

Garden-variety common sense also suggests that neither the internet nor even the proper use of the internet, whether along the lines of crowdsourcing or open data, can provide a universal solution and a little magic to all economic, social, cultural, or political problems. When all that almost everyone can utter are words like *disruption, innovation, networking,* and *collective intelligence,* these order-words no longer make sense because they no longer make a difference.[3]

In a similar vein, the internet is no model either. Wikipedia (since 2001) without a doubt indicates success in the area of collaborative work and the dissemination of knowledge. But must one imitate it, however, for projects and in contexts that are different from that of an online encyclopedia? One can say the same of other successful initiatives such as open source software (since 1983) or Creative Commons licenses (since 2001). Wikipedia and free intellectual property are now interdependent and well-established institutions. If you have to copy the Wikipedia community or the "open" community, it should be their ability to conceive nearly from scratch what unique models they themselves needed *for their own projects.*

Here it is 2013, and there is no reason that new, original models cannot be added to these, perhaps looking forward to more ambitious projects. One certainly must make fruitful our technological, legal, and organizational inheritance of the multifaceted sociotechnical movement that has carried the emergence of the algorithmic medium, but why should one conform to any given model?

To wrap up this section on the critical eye, let's look at some trending order-words such as *big data* and *digital humanities.* It is clear that the vast amounts of publicly available data call for a concerted effort to extract from them a maximum amount of useful information. But the proponents of big data maintain the epistemological illusion that they can do without theories and that it is possible to extract knowledge thanks to "simple" statistical analyses of the data—as if the selection of data sets, the choice of categories that are applied to them, and the design of the algorithms that process them did not stem from a pragmatic perspective, a particular hypothesis, and, in short, some sort of theory! But can one ask engineers or journalists, as well intentioned as they may be, to explain humanities theories, when researchers in the humanities themselves provide so few of them, which are so poor, so simplistic, or so limited to a given situation?

This brings me to the contemporary craze for digital humanities. Efforts to edit and provide access to humanities data, process these data with big data tools, and organize communities of researchers around this processing are surely commendable. Alas, for the moment, I see no substantial work being done to solve the huge problems of disciplinary fragmentation, testing assumptions, and theoretical hyperlocality that prevent the humanities from emerging from their epistemological Middle Ages. The technological tools are not enough! When will the humanities rid themselves of the postmodernist spell that bars access to scientific knowledge and a universal open dialogue? Why do so many researchers, although very talented, confine themselves to politico-economic denunciations, to the protection or attack of such or such an identity, or to disciplinary confinement? It will no doubt be necessary to mobilize new algorithmic tools (the *digital* part), but it will above all be necessary for the humanities community to find new meaning in its mission (the *humanities* part).

I started out by saying that we needed to open both of our eyes to understand and act: the critical eye and the visionary eye. The critical eye dissolves intellectual idols that obstruct our cognitive field. The visionary eye discerns new problems, explores futures hidden in the fog of the future, and creates. This is where a *design* perspective fits, so well evoked by Stéphane Vial in chapter 6 of this book. But before one thinks about creating, one first has to discern. Humanity is the only animal species that manipulates symbols, and this uniqueness has given it access to reflective consciousness, culture, and history. As soon as a new world of communication—a universe that is obviously the fruit of its own activity—increases and modifies its capacity for symbolic manipulation, it is the very being of humanity—its ontological uniqueness—that is called to rebuild itself. But the algorithmic medium gathers and, in a ubiquitous mode, interconnects with the digital data streams emitted by our activities as well as armies of symbolic automata that transform and offer up these data. Since the twentieth century, a few visionaries have dared to face the anthropological mutation that this new regime of symbolic manipulation suggests. It is now time for the technosocial conditions of the mutation in progress, the gaping problems that it poses, and the incredible opportunities it opens to us to be taken into account head-on by the humanities research community.

As this book shows clearly, the "digital revolution" is not so much about appearances or what is observable, which journalists are professionally constrained to; it is more about the "organizing system" of our perceptions, thoughts, and relationships—about their new mode of appearance, the cognitive factory they are, their "nurturing nature." So let's open our visionary eye, walk through the mirror, and start exploring this transcendental and historical change: the emergence of a new episteme. It's clear to me, as I believe it is to Stéphane Vial and many others, that this change is the work of humans, that it is not finished, and that it offers many possible ways of intervening and inflecting creatively. But in order to realize the most fertile potential of our historical and cultural evolution, we still have to envision them and give ourselves not only the technological means to achieve them, but also symbolic, theoretical, and organizational means as well.

There are certainly some demands to live up to: cultural, economic, technological, and existential requirements. Cultural: to not disregard local traditions or traditions transmitted by past generations; to respect the treasures of knowledge and wisdom contained in the living institutions. Economic: regardless of the chosen options (for example, public, private, commercial, noncommercial), our projects must be viable. Technological: let's get acquainted with the algorithms, with their computability and their complexity. Existential: designing experiences must take into account the bodily, relational, emotional, and aesthetic existence of human beings using technologically interactive devices. Once these requirements are met, creative freedom has no limit.

For my part, I believe that the most promising direction for us to evolve in is to make *a reflexive, collective intelligence leap* within a general framework for human development.[4] This cultural and cognitive project rests on a technosymbolic system of my invention: an algorithmic metalanguage (information economy metalanguage, IEML), which self-translates into all languages and provides the humanities with a powerful tool for categorization and theoretical explanation. This project does not exclude any other project. I invite us to think and discuss in open universality. My philosophy, like Stéphane Vial's, welcomes the emergence, duration, and evolution of creative and interpretive uniqueness to be distinct and interdependent as well as competitive and cooperative.

It seems we have forgotten why we built the algorithmic medium. Was it to become millionaires? Was it to finally reveal to oppressed peoples the "social media marketing" they anticipated with such hope? Was it for everyone, from schoolchildren to the most powerful armies, not to mention companies and political parties, to monitor, slander, and destroy enemies more effectively? Stéphane Vial reminds us of what we were after, what we are still after—goals that seem to slink away as we approach them and yet orient our path: *a revelation in human subjects.*

Introduction: What Is the Digital Revolution Revolutionizing?

> Reality is never "what we might believe it to be": it is always what we ought to have thought.
>
> —Gaston Bachelard, *The Formation of the Scientific Mind*

The New Technological Mind

Since the appearance of the first computers in the 1940s, our civilization has been involved in a profound upheaval, which we understand today to be not simply technological. At the beginning in the 1950s, it was only a matter of computerizing our productive systems to improve performance thanks to the computing power of supercomputers, those machines weighing several tons that occupied vast rooms. Then we understood that such machines could be accessible to all and render services to everyone with the microcomputers of the 1970s—that a handful of "computer nerds" and other "hobbyists" worked hard to design, produce, and distribute[1]— but especially with the graphical interfaces of the 1980s, which gave these machines their friendly and Dionysian dimension.[2] It was then that the World Wide Web arrived and transformed the Internet—a technology of computer networks that had 213 computers connected in August 1981, in a global cyberspace that had reached 5 billion connected devices by August 2010,[3] and where a real "life on screen"[4] as much as a new and authentic type of "culture" developed in the 1990s.[5] After the burgeoning of Web 2.0 during the 2000s, emblematized by social networks such as Facebook and Twitter, then with the meteoric rise of mobile technologies and the internet of things, of 3D printers and big data, today each of us feels the incredible magnitude of the phenomenon because each of us is implicated.

Faced with these transformations, it is not simply rhetorical to speak of a "digital revolution," a phrase that today has become a social fact. But what makes it possible to actually speak of a "revolution"? What makes these changes in digital technologies worthy of being considered "revolutionary"? What is turned upside down or upset, what reforms or transforms itself, moves or replaces itself, in what we call the digital revolution? In short, what is the digital revolution revolutionizing?

These are the questions we try to answer in this book by showing that this "digital revolution" is not just a technological event but also a philosophical one. As Bachelard wrote in 1934, "Science in effect creates philosophy."[6] We will see how technology creates philosophy and how digital devices—which all technical devices are in general—materialized theories of reality or reified philosophies of reality. This does not mean, as Gilbert Simondon once pointed out, that "what resides in machines is human reality, human gesture fixed and crystallized into working structures."[7] It means that technical and technological devices are—and have always been—"philosophical machines,"[8] which is to say, conditions of possibility for reality or, better yet, generators of reality. This is what we are calling "ontophanic matrices," that is, a priori structures of perception that are historically specific and culturally variable.

It is appropriate to emphasize with Bernard Darras that "in just twenty years, much of human activity has moved into the digital world, and the development of personal computers, the internet, and mobile telephony have radically changed our relationship to the world."[9] With technology, it was never about anything else: our relationship to the world. That's what we will keep showing: our relationship-to-the-world, as a phenomenological relationship with "things themselves," is fundamentally conditioned by technology and always has been. The digital revolution is not a beginning but one of these "ontophanic" rebirths of which there are few in history. To demonstrate this, we will try to build "a real study of the philosophy of technology"[10] that goes beyond the fascinating seduction (blind technophilia) or respectful fear (easy technophobia) generally associated with the internet and new technologies.

Technology and the Question of Being

Classical philosophy has accustomed us to thinking that our perception of reality results from an interaction between a subject and an object. As if objects and subjects existed in ontological suspension, above the movement of history, detached from the conditions of the century. As if the situation of our being-in-the-world, to recall Martin Heidegger's formula, were dissociated from a culture's themes. As if the being-there (*Dasein*) defined something other than being-here-and-now. Metaphysics always liked to step behind this substantialist postulate. It allows it to avoid thinking through the accidental world in which we live, in favor of a general and essential world, cut up in a scholarly way into universal and eternal categories. And too bad if being is always a product of its times. Too bad if the real is always molded by a culture. Priority to substantialist ontology, flowing in the veins of philosophers ever since they emerged from the Cave. Yet Peter Sloterdijk paved a new and fruitful way by showing that the time has come for philosophy to "experiment with a new configuration between ontology and anthropology":

> It is now a matter of realizing that even the apparently irreducible situation of the human being which is called Being-in-the-world and is characterized as existence or standing-out into the clearing of the Being, represents the result of a production in the original meaning of the word.[11]

In other words, a Being is a *poiesis*, that is, an anthropo-technical construction. Worse: existing is the result of a manufacturing. And the technical/technological combined with others factors takes an active part, not to mention a major one. Man is no longer an essence, a separate substance, but a manufactured process, to be constantly made. This aspect of Sloterdijk's thought reinforces the hypotheses of this book. Philosophy no longer has anything to do with the ontology of substance. It is time for it to consent to becoming onto-anthropological by taking in the empirical results of the social sciences. Maybe philosophy will then understand that the concept of the "technical" and the "technological" are themselves surpassed, because each carries the substantialist idea that it would be a kingdom of objects next to a world of subjects. Philosophers of technology, alas, still often nurture this illusion by stubbornly talking about "technical/technological objects," as if only objects were technical/technological. However, not only are "the products of material culture ... not passive objects, but [they are also] mediators of beliefs, representations, habits and agencies"[12] (hence the interest in speaking of *material* culture rather than technical culture). It is being itself that is technical, that is material. The technical/technological lies not only in objects but in subjects as well.

The digital revolution then functions as a digital revelation. It makes us discover that the questions of being and of technology are one and same question because, though it may always have been true, it has not always been visible. It took digital technologies to bring us "perceptions of an unknown world" to grasp this, just as modern physics brought us "messages of a unknown world."[13] These unheard-of perceptions, which we have been trying since the 1970s to integrate more or less well into the realm of our phenomenological habits, are those emanating from digital devices. Totally breaking with established perceptual culture, these new perceptions give access to beings we had never seen before and whose reality we are struggling to believe. These beings emerge from our screens and our interfaces and, not without causing some perceptual vertigo, upset the very idea for us of what is real. As French psychologist Yann Leroux points out, the "Internet forces us to reflect on what until then, without thinking too much about it, we called *reality*."[14] This question is intensely philosophical. What can one indeed say about the being of this both sensible and intelligible thing that is a menu icon on a digital interface, an avatar on social networks, or a virtual character in a video game? Is it the same thing as a piece of wax? Or is it, rather, a piece of mind matter? Or it is one of those realities we call *virtual*? But what is hiding behind this deceptive term, *virtual*? What is the being of digital beings? And above all, what are they doing to our being? What becomes of our being-in-the-world in the age of digital beings?

This book is a work of philosophical research. It seeks to renew the analysis of technology conceptually in general and in digital technologies in particular. It aims to deconstruct the concept of *virtual* "in all its weighty clumsiness."[15] Although it is originally philosophical, it is not useful to philosophically grasp the nature of the digital phenomenon. Twenty years of daily use of interfaces show us that the virtual dimension is but one of the experiences we have with digital devices. We need new concepts—concepts better able to grasp the complexity of the digital phenomenon and to illuminate more deeply the meaning of what we experience when facing interfaces. That is why this book proposes to introduce the general concept of *ontophany* and thereby to investigate the fact of the digital through the lens of phenomenology. In general, this book is a meditation on technological materiality and perception. The digital is studied here as a *phenomenon*, that is, as what makes an appearance and gives itself over to the subject, through interfaces and thanks to them. That's why, beyond the felicitous euphony that links it to the tradition of existential phenomenology, the title of this book makes metonymic use of the word *screen*, used here to refer to any digital interface.

1 Technology as a System

> After all, there is more in technics itself than in all of what philosophies *du jour* expressed about it.
> —Jean-Pierre Séris, *La Technique* (1994)

At the dawn of the twenty-first century, technology looks more than ever before like a phenomenon of infinite complexity and elusive variety. Alone, the enormous accumulation of tools and processes, know-how and inventions, machines and artifacts create a vertiginous and, as it were, disproportionate whole, whose history coincides with the history of civilization itself. It seems all the more difficult to apprehend a phenomenon-unit that it is defined from the outset as multiple: the very "term itself, `techniques,' is usually used in the plural—there are textile techniques and iron and steel making techniques,"[1] and today we can add digital technologies. Furthermore, each technology considered on its own is also in fact just a "technical combination."[2] It can be broken down into *operations*, which require *tools* that are applied to *materials*, which are transformed by means of *energy*. In other words, from the most basic stage (cutting down trees) to the most complex stage (nuclear technology), technology is always an amalgam of combinations that involves a variety of factors. As Bertrand Gille clearly showed, however, in his monumental *Histoire des techniques*, to which we are indebted, there are several levels of technological combination, whose complexity is growing and whose analysis reveals a certain homogeneity under the name of "technological system" at the heart of the technological phenomenon.

What Is a "Technological System"?

The first level of technical combination is the one we observe at the stage of tools and machines, considered "unitary combinations" capable of structuring materials to accomplish a given task. That is the "technical structure" level: an elementary structure such as a saw, for example, or a complex structure such as a loom, or, in the electronic age, a transistor.

The second level is one that takes shape when several tributary technologies work together to achieve something technically complex, such as industrially casting iron using blast-furnace technology, which involves extracting ore, combusting coal, burning coke, reinforcing furnaces, as well as elevation mechanisms, air blasting, and more. This is the *technical ensemble* level whose "every part is essential to the required result."[3] In the digital age, such an example exists in the industrial manufacturing of microprocessors, which are an essential component of microcomputers. That process includes electrometallurgy enabling the silicon production the Valley gets its name from; microelectronics to connect millions of semiconductors, or *transistors*, on an integrated circuit or *silicon chip*; and computer science, understood as the "science or technology that automates the processing of information."[4]

The third level is reached when several *technological ensembles* are combined in turn to constitute a coherent segment of production dedicated to manufacturing a particular type of product for the end user. This is the level of "technical concatenation" that Bertrand Gille defines as a "combination of technic ensembles which is designed to produce the final product."[5] A good example of this is in the textile industry, conceived as a production complex of clothing, linens, or composite materials, which brings into its complex such different *ensembles* as the transformation of natural or synthetic fibers, spinning, weaving, dyeing, and bleaching. This illustrates the "hierarchy of diverse techniques which contribute to the working of the technical complex" that shapes this concatenation.[6] In the digital age, one can take the example of the computer industry itself: it produces microcomputers, networked devices, mobile terminals, software, applications, and data processing algorithms, although, under the effect of so-called bottom-up innovation or innovation by use,[7] such production is not merely industrial. Still, the technological complex according to Bertrand Gille comes closer to what we call a *sector* in economics: a single production set bringing together similar product families and coherent trade families.

And then we need to describe the fourth level, which encompasses the preceding ones and goes beyond them: that of general coherence, which, including the different previous technological combination levels, makes up all of the technologies of an era when these, reaching their greatest maturity at the same time, become interdependent and organize themselves into a vast homogeneous and characteristic ensemble:

> In general all techniques are dependent upon the others, and this necessarily requires a certain coherence: the coherence within the structures, ensembles and series constitute what could be called a *technical [technological] system*.[8]

When all levels of technical/technological combination reach a balance, which may require several centuries, you get a "viable" technological system, which stands as a model and which, after its heyday, endures until new innovations lead to its being superseded. But to reach this balance, "the technical ensembles have [to reach] a common level."[9] The most significant examples are those of the premechanical technological system that developed in the West from the fifteenth century in the inventive proliferation of "engineers of the Renaissance"[10] and the mechanical technological system (the first industrial technological system), which matured around 1850 through the joint development of metal, steam, and coal technologies that triggered the first Industrial Revolution and the mechanization of production.[11]

From then on, the technological system is the highest-level, most complex technological concatenation that can be observed in any society in that it aggregates in an orderly and consistent manner all other, lower, levels of technological combinations. It is the embodied social form of technological phenomena considered in their entirety and whose concrete organization it is able to describe. As such, it is a fundamental social structure and is implicated in what makes up the identity of an era. It helps reduce the diversity of technological phenomena to their combined essential and historically verifiable expression. That's why this is the historian's privileged object of study: "Those enamoured with chronological divisions are thus able to thus define history in terms of a series of different successive technical [technological] systems."[12] This is what Bertrand Gille himself works on, establishing in the same gesture a systemic approach to the history of technology, whose principles and results we borrow here because we have to base the philosophy of technology on the history of technologies, just as epistemology, since Bachelard, is based on the history of science. To never dissociate the philosophical purpose from historical materiality is the only way to avoid the arbitrariness of ideology. A philosopher of technologies, according to Simondon, must first be a "mechanologist."

Against the *Technical System* and Techno-Fetishism

When Bertrand Gille published *The History of Techniques* in 1978, it followed the well-known book by Jacques Ellul about technical systems published in 1977.[13] In this tormented work, which is primarily a response to the anxieties of his time, Ellul also offers an analysis of the technological in terms of system: "Technology is not content with *being*, or in our world, with being the *principal or determining factor*, it became a System," he says.[14] Like Gille, Ellul summons the concept of a system in the sense of the interdependence of technologies: "But it is a system in that each technological factor (a certain machine, for instance), is first linked to, connected with, dependent on, the ensemble of other technological factors before it relates to nontechnological elements.[15]

Yet then Ellul gives it a wholly different meaning. Unlike Gille, who shows that technology makes systems in any age, Ellul defends the idea that technosystematicity somehow characterizes our times, that it would in a way be a symptom of our age. This bias, which is hardly verifiable historically, can be explained by the ideological struggle that underlies the approach and which, despite the author's talent, often borders on caricature: "There is a system just as one can say that cancer is a system."[16] From then on, Ellul uses "technological system" to refer to the "conjunction between a technological phenomenon and technological progress,"[17] the former designating for him the generalized rational imperative of optimal efficiency, the latter the capacity technologies have to produce their own change autonomously (self-growth)—as if it had some of that "formative power"[18] that Kant has shown is the prerogative of living beings alone. From this perspective, the technosystem "does not leave the social body intact" and invades all spheres of existence: "Total technization occurs when every aspect of human life is subjected to control and manipulation, to experimentation and observation, so that a demonstrable efficiency is achieved everywhere.[19]

And since single misfortunes never happen alone, Ellul accuses the technological of being greatly responsible for this depletion of meaning in an age when signs are consumed, which Jean Baudrillard also denounced at the time.[20] Thus, Ellul continues, "it is the technological [that] erases the very principle of reality" for "it is technology that brings out the non-reality which is taken for reality (consumer goods or political activity). Technology does this by its own process of distribution, the image—and it is technology that 'hides itself'[21] (behind that luminous play of appearances."[22] Of course. The denial is so crude that it is impossible to avoid interpreting it. Endowed with intentions, the technological is presented as an abstract person pursuing his own ends, autonomously and independently of humans. What a disappointment to see this great thinker of the technological falling into such a pitfall, protesting throughout to defend himself and ending up betraying himself a little more, even though Gilbert Simondon had warned very early on that

> an educated man would never dare to speak of objects or figures painted on a canvas as genuine realities, of having interiority, good or ill will. However, this same man speaks of machines that are threatening man as if he attributed a soul and a separate, autonomous existence to them, conferring on them the use of sentiment and intention toward man.[23]

It must be said that the majority of philosophers of the twentieth century—with the exception of Simondon—have hardly overcome the level of anxiety in their analyses of the technological phenomenon. In 1953, Heidegger sees in it nothing more than a phenomenon of "Enframing,"[24] which definitively sanctions forgetting beings. In 1964, Herbert Marcuse feels that "facing the totalitarian aspects of this society, it is no longer possible to talk about the 'neutrality' of technology" because, according to him, "technological society is a system of domination which operates already in the concept and construction of techniques."[25] In 1968, Jürgen Habermas views it as an "ideology" coupled with science, industrial production, and state technocracy.[26] Also, in 1977 when Ellul presents the technological system as itself an object, whose development would be imposed on man as much as it would be independent of him, he wraps up thirty years of ideological condemnation of technology as responsible at once for the mindlessness of man, capitalist alienation, and disenchantment of the world.

This is what leads the philosophy of technology in the twentieth century to becoming locked in an ethical anxiety centered on the tormented analysis of the unpredictability of technological progress, itself condemned to the irresponsibility of a process without a subject.[27] In 1984 in *Le Signe et la technique*, a work prefaced by Jacques Ellul, Gilbert Hottois, who popularized the word *technoscience*, supported this idea that technology would bend to a "blind autonomous growth" that would threaten the very possibility of ethics by consecrating the nonethics of technological progress and endowing it thereby with "black transcendence." Even if, in fighting the technoscientific universe, Hottois does not rise to Ellul's level, his words sound like the "watchwords of theologians"[28] and more than ever point to how correct Jean-Pierre Séris's strong remarks are: "Thinking [they are] defending paradoxes when they are trivializing untruths, ... philosophers seem to have found their greatest common denominator in the denunciation of technology."[29]

In this (very) ideological perspective inflected by Jacques Ellul, whose works had international influence, the notion of a *system* seems to be used merely to convey scholarly authority on a fantastical reality that is far removed from reality. We are looking at what one has to call, following the concept Marx invoked regarding commodities, a *fetishism of technology*. That means the tendency to believe that technology is a thing in itself, endowed with an abstract will, which guides the course of human events by pursuing its own ends—as a process without a subject. Dare we say, this is a form of magical thinking, rationalized after the fact. Brilliantly intellectualized as it is, this fetishism of technology is nothing other that the expression of an anxious imaginary built atop the anxiety of losing control of industrial society. So everything happens as if philosophers in this context were the only ones concerned about it and bringing their concerns to the public, where they are sure to find minds that, with their own fears of technology, are waiting to be confirmed by some great philosopher. Being tough on technology becomes the sole way of warding off anxiety and reveals an inability to analyze the technological phenomenon in an objective and reasoned way. It is what Jean-Pierre Séris calls *misotechnology*, this modern hatred of (technological) reason that dominates the thought of twentieth-century philosophers:

> Contemporary technophobia reverberates widely through the discourse of those who profess philosophy, in a form reminiscent of the "misology" or hatred of reason, against which Plato in *Phaedo* and Kant in *Groundwork for the Metaphysic of Morals* warn disappointed lovers of the logos. "Misotechnology" is the modern

form of "misology." It tells us about the philosophy of the day, but fails to truly educate us about technology.[30]

One cannot say it any better: the *technical system* is nothing more than the means of subsuming an ordinary anxiety under a term of art and for which the concept seeks to be a clumsy remedy in the era of the "disenchantment of the world" (Max Weber). The new philosophy of technology that needs to be undertaken must take an entirely other path, in keeping with the objective reality of the technological phenomenon, as manifested by the history of technology and the field of practice of design.

Technology as a Cultural Value: The Lesson of Design

Philosophy owes a lot to cultures outside itself, be they scientific culture, artistic culture, or political culture, to name only the most classic. This is so true that Georges Canguilhem made the attachment of philosophy to an outside the very condition that gives it an inside: "Philosophy is reflection for which all unknown material is good, and we would gladly say, for which all good material must be unknown," he writes.[31] However, Gilbert Simondon, alone against all others, showed very early on that this exploration of philosophy in cultures outside it was always undertaken in such a way as to arbitrarily exclude technological culture on the dubious grounds that it would precisely not be a culture but simply a host of instruments without symbolic substance:

> Culture is unbalanced because it recognizes certain objects, like the aesthetic object, and granting them citizenship in the world of significations, while it banishes other objects (in particular the technical objects), into a structureless world of things that have no signification, but only a use, a utility function.[32]

This cultural castration of technology by philosophy's superego, so to speak, largely explains the existence and success of technophobic twentieth-century ideologies, to which we owe the growth of this deplorable phenomenon: "Culture has constituted itself as a defense system against technics."[33] Instead Simondon insists that what resides in machines is not a blind and abstract rationality, the instrument of inevitable alienation; it's the stuff of "human reality, human gesture fixed and crystallized into working structures"[34] in the same way as in works of art, scientific theories, or political action. For those of us who measure the deep humanity of the digital age, this axiom is obvious, but the "awareness of the meanings of technical objects,"[35] which Simondon wished for in 1958, took a long time to enter minds and seems to still be struggling to do so. Thirty years later, François Dagognet is still proclaiming the same, as if the idea had still not obtained: "An object is 'a total social fact': 'philosopher-semiologists' must learn to read it, decrypt what culture dwells there, whether on its shell or in its only lines."[36]

It must be admitted that French philosophers' efforts in this direction remain tentative—even if one must acknowledge the remarkable research of the Lyon school with the work of a François Dagognet, a Jean-Claude Beaune, a Daniel Parrochia, or a few sparkling exceptions, like Jean-Pierre Séris. As Gilbert Hottois points out, "In France one encounters almost exclusively philosophers who have dealt with technology only occasionally,"[37] to which Canguilhem would add, "attentive as they have been, above all, to the philosophy of science."[38]

While a coming to awareness has begun, we owe it less to philosophers than to institutions, whose openness to technological culture dates back to the beginning of the 1970s with the establishment of the Industrial Creation Center at the Centre Georges Pompidou in Paris. Jean-Pierre Séris notes, "Long excluded from museums by the nineteenth century, mechanized production is finally recognized not only as a document, but as an art in its own right."[39] In the following decades, technological products were more integrated into museum programming, eventually giving birth to dedicated institutions like the Cité des sciences et de l'industrie, created in 1986, or bringing high-tech products into high-culture arenas, such as Game Story: Une histoire du jeu vidéo (Game Story: The Story of Video Games), an exhibit at the Grand Palais in 2011.

Now, when, under the effect of digital technologies, we don't know the full impact of technologies, it's time for philosophy to finally demonstrate awareness by accepting that technological culture as an outside can teach its inside as it does artistic culture or scientific culture. It is time for the philosophical unconscious to stop pushing away technological culture. In order to do so, a new gateway to technical culture is opening up. Indeed, approaching technology using the benchmark of science exclusively or ethics exclusively, the twentieth century's technophobic ideologies have disregarded a major cultural event of their time: the alliance of technology and art.

From this alliance a new industrial culture was born that spread throughout the entire century and that we call design. In 1907, when architect Peter Behrens, a member of the German Werkbund (an organization of artists committed to the applied arts), became the artistic director of the Allgemeine Elektricitäts Gesellschaft (AEG), an electrical engineering company whose products, image, brand, logo, letterhead, factories, and workers' housing he designed, it was indeed *design* that was born. Henry Cole first came up with the word *design* in 1849 in the inaugural issue of the *Journal of Design and Manufactures* and defined it as a way to "wed great art and mechanical skill."[40] Similarly, when Raymond Loewy, starting in the 1930s, redrew the whole panoply of American consumer objects from the locomotive to the pencil, giving them smooth, rounded shapes and streamlined features and style—this too is *design*, which has its first hour of glory. It is what Americans call industrial design and what the French call "industrial aesthetics."[41] Throughout the century, philosophers nevertheless ignored it, without realizing that a whole new discipline was born, whose history today is well established with its professional practices, work methods, educational institutions across the globe, and recognized major players known to all.[42]

But what is born with design? A new culture that mixes art, technology, industry, engineering, science, philosophy, and social science and is driven by the hope of innovation in the service of humanity. It bears the creative alliance of many once disjointed disciplines that crystallize in an intellectual culture located at the crossroads of thought and action. This culture has gradually reorganized industrial processes by putting humanity at the heart of design and production even when sometimes at the cost of some commercial slippage. That is what Jean-Pierre Séris means when he writes that "postmodern design is distinguished by ... the forceful return of 'meaning.'"[43] This radically changes the nature and form of technical phenomena: "The beautiful, everyone will admit it, moved to the side of industrial technique; it moved, now freed of its constraints, away from art."[44]

Contemporary technologies no longer have anything to do with the social ugliness of nineteenth-century blast furnaces, the misery of the mining facilities of Zola's *Germinal*, or the barbarism of Nazi Germany's gas chambers, whose shadow has long weighed on philosophy. At the edge of third millennium, at the heart of what Bernard Stiegler called the "hyperindustrial era,"[45] contemporary technology resides more with the elegance and effectiveness of Apple products, the subtle aesthetics of the Ricola brand factories (such as the one in Mulhouse that architects Herzog & De Meuron built in the 1990s), and the lightness of Vélib' bicycles, Paris's self-service bikes whose parking stations designer Patrick Jouin designed.

It is impossible under these conditions to persevere with technophobic stubbornness. The appearance of design upsets the order of cultural values by integrating technological cultures and related cultures into a "world of meanings." Industrial creative genius, moreover, has become its own full-fledged area of human genius, worthy of inspiring philosophers at least as much as artistic, scientific, or political genius. "Technology," Jean-Pierre Séris reminds us, "in its ever close and constant interaction with all facets of the sciences, is an intellectual activity of as high a level as science."[46] For genius does not discriminate; it moves in where a place is available—in medieval theology as in the painting of the Renaissance, in modern physics as in the computer industry. The genius of Galileo and Marcel Duchamp deserves to be in the pantheon of intelligence as much as Richard Stallman's genius or that of Steve Jobs. Also, just as is art, philosophers take into account what artists say in their analyses of artistic practice, and just as philosophers of science take into account the discourse of scientists, so too must philosophers of technology take into account the discourse of industrialists, engineers, designers, and innovators in the hope of understanding technological reality.

Let's learn from one of the greatest industrialist creators of our time, Apple's founder, Steve Jobs, whose recent and brutal disappearance accentuates his legacy: "Hollywood and content industries imagine that technology is something you buy. They do not understand the 'creativity' element of technology."[47]

What is true for Hollywood is also true for many philosophers: the element of creativity is not one they traditionally associate with technology; that is why they did not notice the birth of design (which is based precisely on this connection) and have no means of escaping trans- or posthumanist scarecrows. As far as this goes, Steve Jobs was prolonging Henry Cole's inspiration, giving it a magnitude that the latter could not have hoped for. Witness the success of Apple products and the "philosophy of technology" that inspires them: "Technology alone is not enough. It's technology married with liberal arts, married with the humanities, that yields the results that make our hearts sing."[48]

One couldn't stand further from a Heidegger's "Enframing" or an Ellul's "technical system." To be honest, we are close to industrial poetry, as evidenced by this very personal but edifying confidence:

> Remember that the 1960s happened in the early 1970s, right, so you have to remember that, and that's sort of when I came of age. So I saw a lot of this, and to me the spark of that was that there was something beyond sort of what you see every day. It's the same thing that causes people to want to be poets instead of bankers. And I think that's a wonderful thing. And I think that that same spirit can be put into products, and those products can be manufactured and given to people and they can sense that spirit.[49]

Putting that same spirit into manufactured products means introducing marvelousness into commercial objects. That is precisely the artistic idea of (industrial) technology, which governs decisions made by Apple, the foremost company in terms of global market capitalization and a leader in technological innovation. Can the philosophy of technology still ignore it?

Some people would not fail to see in these statements the slogans of an electronics dealer paying attention only to the indisputable marketing effects of such remarks. But let us try the (risky but heuristic) assumption that these words are sincere and that the fact of selling a product, any more than selling canvases or patenting inventions, does not prevent one from having ideals.. In Steve Jobs's case, it's even very much the opposite: Apple products are the rather successful embodiments of his personal ideals, to the point that the history of the company and that of its founder are intimately intertwined,[50] which brings to mind Bergson's "indefinable resemblance ... which one sometimes [finds] between the artist and his work."[51]

There is no doubt that technology can be a purveyor of values and that, thanks to design engineering, it deserves at best the label of culture. As Jean-Pierre Séris says, "It is the technological world itself which carries meaning."[52]

But the most remarkable thing is that such a phenomenon can be verified on the margins of the industrial world itself. The free software movement that computer scientist Richard Stallman started in 1985 with the creation of the Free Software Foundation is a majestic illustration of this. By suggesting that programmers around the world affix to the fruit of their labor a free "copyleft" license, outside any patent (or copyright) logic and outside any industrial or commercial framework, Richard Stallman launched a moral and legal revolution in the IT sector, which eventually influenced all of society. When one invokes the free software movement, one immediately thinks of its values of generosity and sharing, solidarity and exchange, insofar as they can contribute to building a better world (and, why not, inspiring a new economy). Wikipedia's success is one of the most striking examples, but it is far from alone. In a few years, the spirit of open source penetrated all areas of design, reshuffling and reorganizing the logic around the values of community sharing and contributory collaboration. Lawrence Lessig, a law professor at Harvard Law School, thus said of Richard Stallman:

> Every generation has its philosopher—a writer or an artist who captures the imagination of a time. Sometimes these philosophers are recognized as such; often it takes generations before the connection is made real. But recognized or not, a time gets marked by the people who speak its ideals, whether in the whisper of a poem, or the blast of a political movement. Our generation has a philosopher. He is not an artist, or a professional writer. He is a programmer.[53]

Who would have believed fifty years ago that the cultural utopias of the twenty-first century would be borne by technologists? Technology is no longer the specter of the century. It is able to produce values worthy of educating humanity and society. Philosophical culture must not only take note of it but also draw a lesson from it: the lesson of enthusiasm— "enthusiasm" here meaning an impassioned confidence in the future and in creative unpredictability, as well as the possibility that follows from exercising a virtuous influence over that future, against the grain of how we typically educate philosophical minds, namely, excessively worshipping historical studies, which make philosophers, according to Nietzsche, "wandering encyclopedias" broken from "the habit of not taking real things seriously any more."[54] This is what usually leads philosophers to adopt a consistently suspicious attitude toward anything new, in particular technology. What Nietzsche said about historians thus applies to philosophers as well: "The historian looks backward; eventually he also *believes* backward."[55] Conversely, industry creatives can say, "If you want to live your life in a creative way, as an artist, you have to not look back too much. You have to be willing to take whatever you've done and whoever you were and throw them away."[56]

Far from being a value invoking naiveté or idolatry, as haters of technology like to say to find comfort in their technophobia, enthusiasm carries the values of innovation and delight. The only risk it takes is that of the success to which it leads us. In the digital revolution of our time, technological progress once again bears hope and utopia.

The "Technological System" in the Age of "Technology"

Under the effect of industrialization, the fantastic acceleration of technological progress since the end of the eighteenth century led all Western societies to live to the beat of "technology," to the great sorrow of those who criticize its misuse, against the grain of its earlier meanings. Yet if the word *technology* has had better luck than *technics*, it is not because of any Anglo fashion effect or a tendency to grant more value to the most scientifically advanced techniques, as Jean-Pierre Séris maintains,[57] but because modern "technics," such as those related to nuclear energy, chemistry, or microelectronics, are no longer only technics; they are at once technical processes, scientific methods, and innovation logics of marketing and design. They are elaborated in R&D departments where researchers, engineers, and designers work together and function in order to "innovent," a neologism Lucien Sfez proposed, collapsing *invention* and *innovation* "to avoid separating foundational science (that invents) from helping technologies (that innovates)."[58]

This is why the term *technology* is so successful and familiar; and that's why we use it without needing to craft another word, such as *technoscience*, which is too dualistic and excludes the industrial dimension of the phenomenon, and therefore its economic determinants or its creative design dimension. From then on, technology, science, industry, and design, as generators of a new culture, converge as "technology."

The word therefore does not promote in a prescriptive manner any new, contemporaneous, technical values and does not angle toward sophistication;[59] it attests, in a descriptive way, to the new factual configuration in which we live: *technics alone no longer exists; it is a converged phenomenon*. The alliance of technics and art, which gave birth to design, is the last step of this convergence, which was first observed in the alliance of technology and science. From this point of view, one can say that technology convergence is what accurately characterizes the technics of our time and confers upon it this transcendence so dear to those who so mistrust technology, according to which technology is now practiced and accomplished without us and outside us.

In a world where accumulated technical knowledge is colossal, it is true that we are paradoxically freed in our daily lives from mastering the slightest technical know-how:

> At least in everyday life, a technical object is characterized by the conjunction of two features: the very elaborate character of its construction and operation, and the convenience of its use, reducing to zero the competence required of the user. Microcomputers, become convivial by the grace of their user-friendly software, are a striking example of this phenomenon.[60]

In other words, we live among and use very elaborate technological objects about which we have no particular knowledge. "Technology, from this vantage point, is the name we give technics of which we feel dispossessed," writes Jean-Pierre Séris.[61] But why see in this transcendence a dispossession? Technology is no more transcendent to humanity than science is or art is. Science is also done without us and outside us, without our having any scientific knowledge and without being scientists ourselves. It's the same for art: we are not all artists. Yet we do not feel dispossessed of science or art. Why would we feel that way of technology? Driving a car without knowing how it works technically or using the computing power of a computer without knowing what is at its heart constitutes a freeing from technology rather than a dispossession of technology. Let's never forget that the technological object, as François Dagognet reminds us so well, is above all "that without which we have no power":

> We know all too well that our fingers do not cut, our nails break, but a knife blade replaces our too soft tissues to advantage. Objects in general therefore constitute our operational nature, the that-without-which we are otherwise without power. Glass captures, divides, and preserves liquids that our hands can not hold. Same goes for the garment that covers and protects us, just as it differentiates us.[62]

One could add: a bicycle carries our body at a speed our feet are incapable of, and a computer performs logical operations with a rigor and accuracy that our brain can only envy. If the technical and technological are out of reach, it is to meet our needs all the more easily. The transcendence of the technical and technological is nothing other than the condition of their immanence. We had rather leave transcendence to theologians and followers of transhumanist faith in order to speak, for our part, about technological convergence.

Nevertheless, this concept should not be confused with the technical combination defined with Bertrand Gille. Technological convergence is vertical: it corresponds to the homogenization of the phenomena of technics, industry, science, and design insofar as they work together to form one and the same complex. Technological combination, though, is horizontal: it corresponds to the aggregation of technical/technological facts organized into different combinatorial levels—structure, whole, sector—in order to form a coherent technological system. Technological convergence as a feature of time does not detract from technical systematicity as a fact of history. Quite to the contrary, it strengthens it because technologies are systems not only among themselves but also in convergence with other components. Technical systematicity as Bertrand Gille conceives of it thus appears as the only acceptable way in the area of technics of entertaining the concept of *system*—not meaning that technology would be a blind and autonomous *technical system*, but in the sense that technology is always a system of relations. Technical systematicity is to be understood as meant by Bertrand Gille and historians of the real, and not in the way Jacques Ellul and the metaphysicians of anxiety mean it: "For the first, indeed, system implies object to know; for the second, system implies object that escapes our catches."[63] The philosophy of technology must be grounded not in the fantasy of ideology, but in the matter of history—and the first lesson in the history of technologies is that technologies have always been systemic:

> Technique ... has always found its effectiveness in its ability to create a system, both a system of material techniques and a system of these with other techniques and with the system of social relations. Far from being a late oddity, it is a recurring feature.[64]

In other words, technological systematicity or the ability of the technological to build a system is a structural fact of history. "A technology is always at the crossroads of several others,"[65] whether it is an automobile or a steam engine, a crank-rod system or a computer, a technology is always a network of interdependencies and relationships of mutual involvement.

2 The Digital Technological System

> Thomas Edison did a lot more to improve the world than Karl Marx and Hindu guru Neem Karoli Baab put together.
> —Steve Jobs

We will not understand anything about the digital revolution as long as we do not locate it back in the sweep of the history of technology, of which it is both a step and an end point. A step, because the so-called digital revolution is never the last of the *tech revolutions*, after the premechanical revolution of the fifteenth century or the mechanized revolution of the eighteenth and nineteenth centuries. An end point because the digital revolution is thunderous and total; in a few decades it has reorganized our entire *technological system*.

The History of Technological Systems and the Mechanization of the World

The tech history of the West is the story of mechanization. As Bertrand Gille has shown, it begins in the Renaissance with wooden machines powered by the force of water and animal traction; skyrockets during the Industrial Revolution with metal machinery propelled by the force of steam and then electric and motor machinery; and, today, untold new developments with networked digital devices that invigorate every part of our lives in the form of computer terminals that accompany us everywhere and work thanks to extensive algorithmic data processing. Viewed at a historical scale as a multisecular dynamic, mechanization is therefore what, in the modern era, catapults Europe, North America, and all Western countries into a logic of progress, putting an end to what stymied old technological systems. With mechanization, technological time accelerates, the life span of technological systems shrinks, innovations happen one after the other, and "tech revolutions" are increasingly frequent. While it took more than fifteen centuries to shake ancient tech systems, it took only three centuries for the premechanical revolution of the Renaissance (the crank-handle system and printing) to reach the first Industrial Revolution (that of coal, the steam engine, and metal); then less than a century to reach the second Industrial Revolution (that of electricity, the combustion engine, and steel); and, finally, barely half a century for the digital revolution to begin (that of computers and networks, the internet, and mass algorithms). In other words, we entered into a time of rapid change a few centuries ago, and everything one asserts (or denounces) today under the aegis of speed is but another degree of accelerated technological creation that we have experienced since the Renaissance.

Given that, the history of technologies can be read as one of technological revolutions, as a succession of faster and faster technological systems. A heuristic analogy between the works of historian Bertrand Gille and epistemologist Thomas Kuhn will help to better understand this. The former described the historical evolution of technological systems; the second is known for his analysis of the dynamics of "scientific revolutions."[1] The analogy is the following: a technological revolution is the historical development of technologies that consists in a change of *technological system* in Bertand Gille's sense, just as a scientific revolution is the historical development of science consisting in a *paradigm shift* as proffered by Thomas Kuhn. In short, the history of any technology until today has but been the history of technological revolutions, and the history of technological revolutions only that of technological systems.

The historical sequence becomes obvious. The first great machine revolution is the premechanical revolution of the Renaissance. The second is the mechanical revolution of the industrial era. And the third is the digital revolution of our age. What all these revolutions have in common is that each introduces a new reign of machines, gradually leading to a complete change of technological system. The history of technological revolutions is precisely one of continuous (and faster) mechanization. In its first historical form, mechanization is about machines.[2] It is about replacing bodily effort and manual operation by machines, that is, motorized or automated metal appliances. In its second form, the "mechanization" is digital. It is about replacing intellectual and cognitive effort using digital machines, which deal with information in a massively automated manner—proof that mechanization (no offense to dictionaries) is not about the machinery but about the process of automation.

The Question of the "Contemporary Technological System"

In 1978, Bertrand Gille was the first to note that the "modern technological system," born of the second Industrial Revolution, was disappearing.[3] Like its predecessors, this system enabled the advent of new machinery, namely, electrical machinery and motorized machines. Like its predecessors, it developed thanks to a limited number of innovations: the production of electricity and the exploitation of oil. And yet as early as the 1970s, it became apparent that major change was underway. Politicians called it a crisis. Enlightened philosophers called it "world changing."[4] The politicians didn't know how to govern it, and the philosophers knew that it would happen inexorably. No one, however, can say when it started, when it will end, or even what will be the end result. But a new "technological revolution" obviously is afoot: we are engaged in a "major movement which is currently taking place within the technological system,"[5] Bertrand Gille points out. The question is what exactly is the "technical [technological] system in its current state."[6] Is there a "contemporary technological system" that is new enough to be distinct from its predecessors? For Gille, there is no doubt: "A new technical [technological] system was indeed created where the most significant factors were already established, forming the coherence which is indispensable within any system."[7]

The problem is that Gille does not have the requisite distance in time to assess the exact nature of the new technological system. With death hovering over him (he died in 1980), he probably wanted to perfect his monumental history by closing it on a description of the contemporary technological system, which, obviously, can be neither exhaustive nor convincing—and he sensed that with great foresight: he admitted that his last chapter would "soon become outdated."[8]

A change is in progress for sure, but in 1978, one could not yet make out its systemic coherence. It is therefore with a certain critical distance that one must consider Bertrand Gille's hypothesis of the "contemporary technological system."[9] According to him, this already constituted new system is based on three main innovations: nuclear energy, new materials, and electronics. To be persuasive, Gille urged us to study the new lifestyles that came with it; they are easily observable among the objects used in a 1970s apartment or kitchen. Indeed, he notes, pestles, sifters, two-stage coffee makers, dishwashing basins, washpots, heavy-duty sewing machines with cast iron frames and pedals, or hand-cranked telephones have disappeared from our homes. Instead, we have electric *household appliances*: washing machines, coffee grinders, toasters, electric beaters, centrifuges, electric knives, mixers, kettles, electric fryers, but also transistor radios, ballpoint pens, pocket calculators, and so on.

No one can deny these observations, which honor the rigor of the historian. However, it does not therefore mean that a new technological system is already in place. Bertrand Gille was mistaken about the nature of new technologies that structure this system. Nuclear energy, for example, as innovative and impressive as it is, belongs more readily to the second industrial technological system, which it improves, than to the new system that is emerging. Ever since the Fukushima disaster in 2011, it even seems to have become a technology of the past that several countries are trying to abandon.

That said, Bertrand Gille's intuition was on target about what he himself called the "electronic revolution."[10] Of course, he did not imagine the scale it would have, but he understood its importance, without alarmist prejudice at a time when, as Jacques Ellul stresses, "man appears incapable of foreseeing anything about the computer's influence on society and humanity."[11] There is, moreover, some irony in the fact that the year before the publication of Gille's *The History of Techniques* in 1977 was when the Apple II had its debut and created the microcomputer market. Gille clearly perceived the systemic potential of computing, even though he did not have the means to draw the necessary conclusions we see a posteriori. He suggested, however, that the novelty and coherence of the new technological system rests on computation:

> The computer has sort of become a symbol of modern civilization. You see it everywhere, in management, industry, accounting, space flights. It makes everyone's job easier, it solves all problems, it threatens public liberties.[12]

The same might be said today about the internet—just as much about the pervasiveness of computers and networks as about the concerns that this ubiquity creates. But what emerges in such sentences, what is pointed out without being named, is precisely the idea of systematicity. When we see a technology *everywhere*, that's a system. Bertrand Gille himself taught us this. So we need to be more Gillean than Gille himself and assert that the computer is none other than the total technological object at the core of the new technological system. It's so obvious that Jacques Ellul himself noticed it and described it as the "end all and be all of informatics":[13]

> In reality, *it is the computer that allows the technological system [système technicien] to definitively establish itself as a system*. First of all it is the computer that allows large subsystems to organize. For example, the urban system can close itself up only because of the urban data banks (Census results, building permits granted, housing already built or under construction, water, telephone, power, transportation, and other networks). Likewise, the air-communication system can function only due to computers, given the complexity, the very rapidly growing number of problems sparked by the multiplication of transportation combined with the technological progress in those areas (it is not only the often-mentioned booking of seats, but also, for example, the permanent relationship of each airplane, at every moment, to every moment with a huge number of control centers on the ground). The computer also makes possible the large accounting units, that is, the infrastructure for an unlimited growth of economic and even administrative organizations. Would it help to recall the importance of the computer as a memory for scientific work? It is the only solution for preventing the researcher and the intellectual from being swamped by documentation.[14]

What was true then is so much more so today. The advent of graphical interfaces in the 1980s, the rise of the internet in the 1990s, the success of Web 2.0 and social networks, the growth of mobile communications in the 2000s, and the advent of big data are some of its most more remarkable manifestations. Seen from the perspective of the history of technology, informatics is the true innovation of our time—the one that is a *system*. It brings along new machinery, that of networked digital machines, which includes large mainframes; microcomputers; web servers; video game consoles; interactive kiosks and mobile terminals; smartphones and tablets and e-readers; connected objects and driverless cars; 3D printers; and more. Not only is the computer a new machine, but it is also a total machine, of which you can ask almost anything or just about, including defeating world chess champion Garry Kasparov (Deep Blue, IBM, 1996–1997).

In other words, the current technological revolution, which Gille and Ellul could only glimpse, is a revolution in informatics and networks, of the internet "and everything which is associated with it: multimedia, computers, informatics, information,"[15] and which we call the digital revolution. It is establishing a new technological system—and we cannot tell just how far it will go, but we all feel its huge effects in our daily lives. Maybe this new revolution is merely in its infancy. Because even if their pace increases a little more each time, technological systems take a long time to reach a point of general coherence. Perhaps the information revolution is only the first step in a larger-scale process that will, historically, lead authentically to a third Industrial Revolution. That is economist Jeremy Rifkin's thesis: "Great economic revolutions occur when new communications technologies converge with new energy systems."[16] According to him, digital technologies, especially the Internet, are now about to merge with renewable energies to create the dynamics of a new world:

> In the coming era, hundreds of millions of people will produce their own green energy in their homes, offices, and factories, and share it with each other in an "energy Internet," just as we now create and share information online. The democratization energy will bring with it a fundamental reordering of human relationships, impacting the very way we conduct business, govern society, educate our children, and engage in civic life. ... The third industrial revolution is the last of the big industrial revolutions and she will lay the groundwork for an era emerging cooperative. ... In the coming half century, the conventional, centralized business operations of the First and Second Industrial Revolutions will increasingly be subsumed by the distributed business practices of the Third Industrial Revolution; and the traditional, hierarchical organization of economic and political power will give way to lateral power organized nodally across society.[17]

These are perhaps the changes that are forthcoming in the new and creative momentum initiated by the digital innovations of twentieth century. It will probably be necessary to wait a few decades before being fully able to describe this new technological system, but we already know that it rests on the digital. The contemporary technological system is a digital technological system.

The Digital Takes Command: The New Technological System

One of the great lessons we learn from studying the history of technology is the rising tendency to delegate to machines. The more time marches on, the more human beings entrust more and more elaborate tasks to machines that are themselves more complex. In the Middle Ages, manual labor was delegated to wooden mechanisms powered by the force of water (the hydraulic saw replaced the manual saw). In the modern era, the work of the whole body, that of an individual working alone or assisted by animal power, was entrusted to metal machinery using steam, gas, or electricity (bicycles replaced walking; locomotives replaced stagecoaches; agricultural tractors replaced plows; the telegraph replaced the horse-driven postal service).

Today, in accordance with the meaning of the history of technology, mechanization continues and inscribes our present in the continuity of a logic reducing feelings to nothing, maintained by a fashion in drama, so that we are experiencing a radical break with the past. Far from having stopped or decreased, mechanization, on the contrary, is conquering new territories that no one had imagined being one day accessible using machines. Michel Volle was one of the first to notice it, in France, in a book titled *L'Économie des nouvelles technologies* (1999): in the eighteenth and nineteenth centuries, "Mechanization made machines take over the physical work associated with production," whereas in the twentieth century, with the spread of computers and information networks, "automation has it take charge of the mental effort associated with production."[18] This shows a logical continuation in the historical process of mechanization: after the mechanization of bodily work comes the computerization of mental work. Steve Jobs himself said, "What the computer is for me [is] the most remarkable tool that we've ever come up with and it's the equivalent of a bicycle for our minds."[19] As for Bertrand Gille, he rightly saw in the birth of information and communication technologies techniques to "transmit thought."[20] This has been widely echoed in science-fiction films since the end of the 1990s in films like *The Matrix* (Larry and Andy Wachowski, 1999) and *eXistenZ* (David Cronenberg, 1999), which portray a world where digital machines take possession of our minds.

So we entered an era of new machinery anchored in delegating the workings of the mind to digital machines, that is, delegating intellectual work, and even mental leisure (video games), to computers—at least to a certain extent: the point where calculating stops. Indeed, in the digital technological system, calculation, which designs an operation or a set of operations involving numbers, is the fundamental operation to which all the other operations are reduced. Twentieth-century man, and even more so twenty-first-century man, is he who delegates computing work to machines, to computers of all shapes and sizes: large systems; microcomputers; consoles; terminals; tablets; smartphones; and so on. The digital technological system indeed engages a real extension of calculation. Writing a text, sending a message, live-chatting, creating a photograph, listening to music, sharing a video, playing a multiuser game, rebroadcasting a TV show, doing cartographic research, controlling factory production, purchasing a product or a service, recruiting an employee, submitting tax returns, managing bank accounts, voting in elections, public speaking: all of this comes under calculation, because all of it is reducible to information that networked computers, which can process large amounts of data, can calculate.

The mechanization of calculation is indeed the new stage of mechanization. It is the fundamental technological innovation of our time and the starting point of this technical revolution we call the digital revolution. We can estimate that computers are to modern times what crank systems were to the Renaissance or steam engines to the first Industrial Revolution. The computer is the disruptive machine that, by adding an extra degree to the historical scale of mechanization, moves us from the second industrial technological system to another whose development is taking place. There is significant evidence of this in the changes of large-scale economic balances. Today, the world's largest market capitalization company is no longer Exxon Mobil, the oil company, or General Electric, the empire Edison founded, but Apple, which secured this title for the first time on Wall Street's stock exchange in August 2011. The digital revolution amounts to the third, great, technological revolution in the West's modern history. From the invention of the computer in the 1940s and the beginnings of computing in the years 1950 to 1960, inflected with IBM's *big systems*, to the extension of microcomputers in the 1970s and 1980s with the first Apple and Microsoft products, not to mention the upheavals of unprecedented growth owing to the internet in the years 1990 to 2000, the digital is deeply transforming the technological system at hand.

According to Michel Volle, a Bertrand Gille reader, the *contemporary technological system* is not that of an alliance among nuclear energy, new materials, and electronics. For him, it emerges entirely from electronics. More precisely, the contemporary technological system (CTS), the first way Volle writes about it in 1999, is "characterized by the synergy of microelectronics, automation, and computing"[21] or "of microelectronics, computing, and robotics."[22] This new technological combination, which is iconic of the second half of the twentieth century, comes into play according to him around 1975 and determines the new industrial world. The resulting type of production is based on automation. Succeeding mechanization, automation is "the fundamental characteristic of the current developed economies:[23] it "seeks to eliminate the mental effort required in production."[24] From there, in the new computerized economy, everything becomes "computer aided" (CA), evidenced as early as the 1980s with the success of acronyms such as CAD (design) and CAM (manufacturing), to which we can add today, among many others, CRM tools (customer relationship management) and, in the Web industry, CMS (content management system), as well as cloud services. When we speak of "everything" becoming computer aided, we are really speaking about everything related to our minds. French researcher Sylvie Leleu-Merviel highlights this:

> What we call the digital age is characterized, at the most trivial level devoid of interpretation, by a nonetheless major phenomenon: the bursting onto the scene of computers for operations in the realms of cognition, data manipulation, knowledge, information, and communications.[25]

And then at the economic level, it is the entire production process, in its cognitive work, that becomes fully computerized and, at the same time, automatic: "Factories are automata controlled by a few people watching screens"[26] that display the behaviors of robotic and autonomous machines; their operation "requires a small amount of work, monitoring, conditioning, and maintaining."[27] The consequences are phenomenal and well known. On the social level, "automation does away with industrial employment as mechanization did away with agricultural employment."[28] And for good reason:

> With automation, qualified people design products and technologies; they provide plans, diagrams, computer programs, instructions, etc. The cost of physical production is negligible compared to the cost of conceptual design. Product distribution requires service jobs.[29]

It's not so much, as in the days of mechanization, that machines replace human beings one by one, and as many times as there are individuals to replace. It's just that there is no longer any need for a large number of individuals to produce, because directly productive jobs are no longer needed.

Alongside the machines in automated factories, only a few maintenance jobs and conditioning are needed any longer.

The "employment crisis" of recent decades is maybe not due to any particular drop in production, any lack of dynamism in this or that sector, or any sort of failure of this or that policy. It is structural and systemic. "It is general,"[30] writes Michel Volle. Today, we still do not know if the digital economy, although it can generate high incomes, is capable of creating a lot of jobs. As Hubert Guillaud emphasizes, "Whereas economic growth is accelerating, job growth is not following suit"; indeed we hear "innovation without jobs" being talked about.[31] From a macrohistorical point of view, if we follow Bertrand Gille's lessons, we can nevertheless say that this crisis is similar to one of the many structural tensions that accompany the establishment of any new technological system over several decades. Sixty years after the invention of the computer, perhaps we are not far from a point of equilibrium. Remember that it took seventy years for the steam engine to come into its final, operational state.

But at a purely economic level, the victory of automation is inevitable because it achieves better profitability by making production a function of fixed costs:

> In the economies of Antiquity (slaves' manual labor) as well as in primitive forms of agriculture, costs were proportional to the quantity produced. In a mechanized economy, initial investments reduce the cost of additional units ("marginal costs"): the average cost decreases with increasing production ("increasing yield"). In an automated economy, only the initial investment costs; the cost of producing additional quantities is practically nil.[32]

In other words, the new, automated technological system "imposes a sort of natural selection," says Michel Volle. In time, its ubiquity is inevitable, just like a law of nature. Companies that become immobilized disappear, and with them the technological system to which they belonged.[33]

With the establishment of this new production system, which relies thoroughly on computers, we are witnessing the emergence of a new technological system. Michel Volle might have called it an *automated technological system*, but he chose at first to keep Bertrand Gille's *contemporary technological system*. However, because all that is contemporary ends up being old one day—which makes labels of a chronological nature obsolete—we prefer to name it a *digital technological system*, a logical concept that has the benefit of pointing to the dependence of the new system on the invention of digital machines and their derivatives. The technological revolution we are living through can then be called *digitization* (of thought), on the model of *mechanization* (of the body), which characterized the first two Industrial Revolutions.

So it's no longer *mechanization* but *digitization* that takes command, to paraphrase Siegfried Giedion's well-known title.[34] We could also call it *computerization*, as Michel Volle does in his 2006 essay, which begins with these words: "Computerization is the most important phenomenon of our times."[35] Volle reuses it elsewhere, in a 2012 contribution, in which he states, "Computerization is the contemporary form of industrialization."[36] That seems accurate. The third, industrial, technological system is based in a first instance on the computerization of our devices. Just as "industrialization did not suppress agriculture, it industrialized it, ... computerization is not suppressing mechanized industry: it is computerizing it."[37] Michel Volle captures its historical movement very well.

It starts in the 1960s with computerizing "time-consuming operations and paperwork: accounting, payroll, invoicing, managing stock portfolios, taking orders." It's the era of large-scale computers and software based on "programming algorithms that provide a result based on the data entered." The movement continues in the 1970s with "database standardization and information systems architecture" in order to avoid double entries and to enable one application to feed another. Then comes the turn of the 1980s, with "the spread of microcomputers and local area networks," which marks a decisive step in the democratization of informatics and lasts into the 1990s' disruptions triggered by the rise of the internet and the Web. Thanks to electronic documentation and messaging, the internet and the Web enable "computerizing a production process by transferring, from one workstation to another, the documentation for the development of a product."[38] To that we can add that the movement continues in the 2000s and 2010s with the rise of Web 2.0 and social networks, mobile devices and multiscreen services, networked things (IoT) and soon 3D printers, which greatly reorganize social practices. Thus, in fifty or sixty years, the computerization of production systems has become thorough and complete:

> Informatics [is no longer] the information system overlaid on management and production systems: intertwining itself with the work of human operators it [insinuates itself] into the intimacy between management and production and [becomes] inseparable from them.[39]

From the culture of an industrial workforce associated with mechanical machines, we switched to an industrial culture of the "working brain" (*cerveau d'oeuvre* is Michel Volle's term) associated with digital machines.

Thus, we are the children of computers. To express this another way, the computer is our "total technological object." By that, we mean a technological object out of which the entire technological system is developed and structured, a bit like Marcel Mauss's "total social fact" summarizes all of a society's institutions.[40] The crank-rod system was the total technological object of the classical technological system. The steam engine was the total technological object of the first industrial technological system. The automobile was the total technological object of the second industrial technological system. The networked computer is the total technological object of the third industrial technological system, the digital technological system. That's why it is an innovation, meaning what Bertrand Gille does from a macrohistorical point of view. According to him, to innovate is not simply to invent, but to invent something that is required at every level of a technological system. In this sense, the computer is *both* an invention *and* an innovation.

Nevertheless, computerization is not enough to describe the digital because what actually confers on the computer its ability to become a total technological object is that it is and is used on a network, which developed in a second stage and has grown during the past twenty years to be what we experience now. Computers have been an operational technology since the 1950s, with visible economic consequences as early as the 1970s; networks for their part appear in the 1960s, with social consequences that surface only in the 1990s thanks to the spectacular rise of the World Wide Web (WWW). Thanks to the generosity of its inventor, Tim Berners-Lee, who had the courage to put his invention in the public domain, the Web turns computers into servers and creates a new world, the one we temporarily called *cyberspace* and that Pierre Lévy defined in the 1990s as "the new medium of communications that arose through the global interconnection of computers."[41] Initially restricted to local area networks, network technologies quickly spread to Web servers, but then also to mobile terminals, tablets, networked things, and more. In this sense, Paul Mathias is correct in pointing out that "the internet is a total phenomenon" that "traverses all strata of life":[42] The internet would therefore be the informational analog of the steam engine, and we will have brought extra cognitive power while extending the horizon of our social and cultural practices.[43]

Its involvement can be seen well beyond the strictly economic dimension in every aspect of societal life: legal (open software and *copyleft*), intellectual (Wikipedia and other participatory services), cultural (peer-to-peer platforms or the Apple Store), leisure (video games played on consoles or the network), social (Facebook and other social networks), politics (blogs and Twitter microblogging), recreational or creative (Instagram, Pinterest, and other networks of interest) or scientific (Google Scholar, JSTOR, Cairn, OpenEdition).

If the computer is the total technological object of our time, it's not only because it has made its way everywhere since the 1950s, but also because it has been connected to so many other networked computers since the 1990s. Computers are everywhere, and wherever they are, they carry everything. With networks, we leave the computer era and truly enter the digital era. As such, the internet must be considered an essential element of the digital technological system. Michel Volle understood this well. In 2006, he referred to the technological combination on which the new system rests as the "fundamental synergy … between microelectronics and software."[44] But more recently, he has specified the nature of this combination by evoking the "synergy of microelectronics, software, and telecommunications networks."[45] Indeed, the fundamental technologies, whose combination defines and structures the digital technological system, are electronics (the physics of components side), computing (the logic of algorithms side), and networks (the reticular connections side). The new technological system is not simply computer based, but networked digital computing, based on the combination of computers and networks. While the computer is the central star of the system, the internet is the orbital structure that allows this star to glow in every place and in every part of the world.

3 The Technological Structures of Perception

> The word *technique* [technology] evokes not a specific form of activity but a particular aspect of all of our activities.
> —Pierre Francastel

At the end of the fifteenth century, the painters of the Renaissance changed our spontaneous way of looking at nature by inventing the landscape, "a form into which perception flows," thanks to which we learned to see in perspective, around a vanishing point, that is, to cut out paintings from nature.[1] At the end of nineteenth century, non-Euclidean geometries upset our intuitive design of space by designing, against the grain of given evidence, other types of spatiality from the one perceived in three dimensions and inherited from Euclid's geometry.[2] At the beginning of the twentieth century, quantum theory challenged the immediate concept we had of the physical world and made it possible to describe the behavior of infinitesimally small worlds thanks to new mathematical concepts.

Such upheavals of our common perception form what we propose to call *phenomenological revolutions*, in the sense that they modify the act of perceiving (for example, nature, space, matter) by affecting our perceptual culture. *Perceptual culture* means all the ways of feeling and imagining the world, inasmuch as they rely on the "habits or skills learned by man as a member of a society."[3] In other words, perception is not just a function of the body or consciousness; it is also a social function, in the sense that it is conditioned by cultural factors. A phenomenological revolution thus occurs when the act of perceiving is affected or modified by an artistic, scientific, or technological innovation. This is what happened in the three cases cited. Observing nature in Ancient Greece and during the Renaissance is not the same thing. Apprehending matter in the classical age or the quantum age is something else yet again. And perceiving reality in the mechanical age or in the digital age is fundamentally different. Digital technologies finally reveal this to us because the technological and the real have always made common cause.

Indeed, since the invention of the computer in the 1940s, the advent of the digital technological system defines the digital revolution at a historical level. But we must now try to understand philosophically what the digital revolution has revolutionized. As with all previous technological revolutions, it is a phenomenological revolution, that is, a revolution of perception; it rattles our perceptual habits of matter and, correlatively, the very idea we have of reality.

To perceive in the digital age is not to perceive new objects as if perception, applied identically to all classes of possible objects, were simply enriched by a new class of objects to which it had only to apply itself as though to any other class. Perceiving in the digital age is to be forced to renegotiate the act of perception itself, in the sense that digital beings force us to forge new perceptions, that is, objects for which we have no habits of perception. This perceptual renegotiation does not come naturally. It requires contemporary subjects to do real phenomenological work to learn to perceive this new category of beings, digital beings, whose phenomenality is unknown and therefore disarming. This at once psychic and social phenomenological work is one that requires each individual to reinvent the act of perception to make it compatible with the particular phenomenality of these beings. It's about learning to perceive digital beings for what they are without metaphysical overstatement or fantasmatic drift, which first involves understanding what they are because, as we will see, perception has never been a function of understanding as much as it is in the age of digital beings.

The digital revolution is therefore not just a historical event in the history of technology; it is also a philosophical event that affects our phenomenological experience of the world, and it is a matter of ontology, or rather ontophany, that is, how beings (*ontos*) appear (*phaino*). But while the non-Euclidean revolution and the quantum revolution were first and foremost intellectual revolutions limited to the restricted circle of scholars able to understand them, the digital revolution is a social revolution that affects all parts of the population. It is a massive event that disrupts the ontophanic experience of hundreds and hundreds of millions of people.

Understanding the digital revolution philosophically is therefore about analyzing what the digital alters in the very structures of perception, which is the only way of understanding the phenomenality of digital beings. But to be able to grasp the nature of digital ontophany, we have to go back to the essentially technological origins of all ontophany. That is the meaning of the hypothesis put forward here. Because of its magnitude, the digital phenomenon makes visible a universal, philosophical feature characteristic of (any) technology, which has remained relatively unnoticed but is essential: that *technology is a structure of perception*. It conditions the way in which reality and beings appear to us. In other words, any ontophany of the world is a technological ontophany.

"Phenomeno-Technology," or Bachelard's Lesson

Ever since Kant, we know that the object is partly constructed by the subject and that our knowledge of the world is less a reflection of it than the combined result of what we receive using perception and of what we produce using reason. According to the author of the *Critique of Pure Reason*, there are indeed structures of sensibility and understanding that set the conditions in which perception and knowledge are possible: it is about the a priori forms of intuition, which space and time are, or the pure concepts of understanding, which the twelve logical categories used to order the diversity of intuitions are. These structures are called transcendental because they are part of the internal organization of our knowledge faculty and, as such, preexist any act of perceiving or knowing as a priori conditions. This means that the act of knowing, just like the act of perceiving, does not come naturally; it is fundamentally overdetermined, which is to say constructed.

This is why constructivism, which owes a lot to the Königsberg philosopher, is most often defined as "Kant's theory that the knowledge of phenomena results from a construction carried out by the subject."[4] We nevertheless have to wait for the twentieth century to watch "constructivist epistemologies" take shape, whose "visible birth," as Jean-Louis Le Moigne says, is in France embodied in the nearly simultaneous publication of two major works *The Construction of Reality in Children* (1937, translated

1954) by Jean Piaget, and *The New Scientific Spirit* (1934, translated 1984) by Gaston Bachelard.[5] The first is the work of a psychologist who puts forward a genetic epistemology according to which "intelligence (and therefore the action of knowing) does not begin with knowledge of the ego, or such things, but with knowledge of their interaction; it is by orienting itself simultaneously towards the two poles of this interaction that it organizes the world by organizing itself."[6] The second is the work of a philosopher of science who supports a historical epistemology according to which, in modern science, "nothing is given" because "everything is constructed."[7] It is Bachelard's epistemological constructivism that interests us here, to the extent that it is based on the decisive concept of *phenomeno-technology*. What does this term mean?

> It is in a brief article of 1931 entitled "Noumenon and microphysics" that Bachelard introduces the concept of *phenomeno-technology* for the first time.[8] Created from scratch, this concept highlights one of the fundamental characteristics of modern science according to which scientific work does not consist in describing phenomena as if they predated the theory that thinks them through, but builds them entirely from scratch thanks to technological devices capable of making them appear and, moreover, of making them exist as phenomena proper:
>
> > The classical division that separated a theory from its application was unaware of this need to incorporate the conditions of application into the very essence of the theory. ... We understand that science *realises* its objects, without ever just finding them ready-made. Phenomenotechnique [phenomeno-technology] *extends* phenomenology. A concept becomes scientific in so far as it is accompanied by a technique that realises.[9]

The example of nuclear physics, dear to Bachelard, enables us to take its full measure. In 1911, Ernest Rutherford hypothesized that at the center of the atom is a *nucleus* concentrating almost its entire mass, with its electrons determining only its size. But because the material of which an atom's nucleus is made is a million billion times denser than ordinary matter (an atomic nucleus is a thousand times smaller than the atom but contains 99.97 percent of its mass), the phenomenal observation of the nucleus seems at first impossible. It is not until 1932 that John Cockcroft and Ernest Walton think of electrically accelerating particles at very high speed to hurl them at the nucleus in order to disintegrate it and thus be able to observe it: it is the birth of the first particle accelerator, which will become the main instrument of nuclear physics.

The atomic nucleus as a scientific reality first existed in a theoretical state thanks to a hypothesis before being made phenomenal thanks to a technical instrument. As Bachelard noted in 1933, that's why "an instrument, in modern science, is truly a reified theorem"[10] in the sense that—still using our example—a particle accelerator is a theory of the atom realized technologically. Thus, "a measuring instrument always ends up being a theory: and the microscope has to be understood as extending the mind rather than the eye."[11] In other words, technical instruments developed by scientific reason are implicated at the core of a theoretical-practical process of active development of phenomena. In *The New Scientific Spirit*, Bachelard explains a little more:

> Phenomena must be selected, filtered, purified, shaped by instruments; indeed it may well be the instruments that produce the phenomenon in the first place. ... And instruments are nothing but theories materialized. The phenomena they produce bear the stamp of theory throughout. ... A truly scientific phenomenology is therefore essentially a phenomeno-technology. Its purpose is to amplify what is revealed beyond appearance. It takes its instruction from construction.[12]

For twenty years, in each of his works, Bachelard hammered this home: "The different ages of a science could be determined by the techniques of its measuring instruments"[13] because a science can only know what its technical equipment actually enables it to do. What's more, scientific realities do not even exist—as phenomena—outside the devices capable of revealing them. To appear, they need a device—hence, the consubstantial link between technology and phenomenon: "The scientific phenomena of contemporary science do not really start happening until we start with machines. So the phenomenon is a machine phenomenon."[14]

To understand this properly, we must go back to the Kantian distinction, repeated by Bachelard, between noumenon and phenomenon. A phenomenon is what I experience through perception. A noumenon, or thing in itself, is what is beyond experiencing. The infinitesimally small world of contemporary physics, this "hidden world that contemporary physicists tell us about,"[15] is a noumenal world in the first instance, that is, inaccessible to experience, because it is above all of a "mathematical essence":

> It is no longer a question, as was constantly repeated during the nineteenth century, of translating into mathematical language the facts that we experience. It is rather, quite to the contrary, about expressing in the language of the common experience a deep reality that has a mathematical meaning before having phenomenal meaning.[16]

To become phenomenal—to become an observable phenomenon of nature—the microphysical world must be made manifest by specially adapted technologies, such as particle accelerators. And for good reason: in the quantum world of atoms, electrons, and particles they are made of, actions are produced that are totally inconceivable on a phenomenal scale, but which from noumenal perspectives are scientifically established. A particle of matter, for example, can exist simultaneously in two places at once and move on its own! Sheilla Jones stresses that even though they understand its functioning from a mathematical point of view, scientists do not know why it works this way.[17] Such behaviors of matter totally disrupt the phenomenal rationality of daily life. In 1935, Erwin Schrödinger also imagined the famous quantum thought experiment of a cat locked in a box, which led to the conclusion that the cat could, from a quantum point of view, be both dead and alive. "Nobody really understands quantum physics," Nobelist Richard Feynman, one of its greatest theorists, is said to have said.

From Bachelard's point of view, the phenomenal reality of the quantum world needs to be built technologically in order to avoid the risk of remaining a world hidden for mathematicians. And without the technological construct capable of making it a phenomenal reality, it does not exist, at least as a *phenomenal reality*—hence the notion of *phenomeno-technology*, whose first formulation in the 1931 article is made fully clear here:

> This noumenology illuminates a phenomeno-technology by which new phenomena are, not simply found, but invented or constructed from scratch. ... Contemporary atomic science is more than description of phenomena, it is a production of phenomena.[18]

Twenty years later, Bachelard had not changed his mind: with contemporary physics, "we have left nature to enter into a factory of phenomena."[19] Science is therefore phenomeno-technological. Rather than discovering phenomena from the outside, it builds them from within, using theories materialized by instruments.

By "phenomeno-technology" we therefore must understand a constructivist technology that manifests phenomena. The major philosophical lesson to learn from this is that *technological constructability is a criterion of phenomenal existence*. It is because a phenomenon is or can be constructed that it can, in modern life, exist as a phenomenon. In other words, technology generates phenomenality. The philosophy of technology does not seem to have taken the measure of such a powerful idea. We can nevertheless infer from this, well beyond the field of science, the foundations of a phenomenological constructivism likely to profoundly alter the gaze we cast—or rather do not cast—on technology.

Technology as Ontophanic Matrix

Whereas scientific experiment designates the rationally and technically constructed apparatus that enables testing a hypothesis in order to produce knowledge, experience broadly understood refers to an order of what can be felt through the senses and, more generally, of what is accessible through perception. Experience is the fact of perception. And perception is the interaction with phenomena. By *phenomenon* one should not understand scientific phenomena in their narrow sense (as in Bachelard's analysis), but in the general sense of worldly phenomena or phenomena of the world, as offered up to the subject. A phenomenon of the world is everything that appears and consequently offers itself up to us to experience and engage with. It is just as much about physical and psychic phenomena as social ones. Hence, experience is the fact of experiencing a phenomenon of the world. Every time I perceive, I am experiencing the world. And since I perceive at every instant, I am at every instant experiencing the world.

Although phenomena of the world include scientific phenomena, the phenomena of the world do not, like scientific ones, require an apparatus. One doesn't need to operate apparatuses to make them appear. Much to the contrary, everything happens as if it were already there. Nonetheless, we want to show that they do not just appear on their own, so to speak, as if it they happened naturally and independently of us. And this is where Bachelard's epistemological constructivism, that of knowledge, founds our phenomenological constructivism, that of perception. Just as knowledge is built in its interaction with the object, perception too is built in its interaction with a phenomenon. Just as Bachelard does not separate science from its conditions of application, we do not separate the perception from its conditions of use. And just as conditions for the application of science are technical, there are conditions of use of perception that are also technical.

The hypothesis is as follows. If technical constructability is a criterion for phenomenal existence, this is true not only for scientific phenomena; all phenomena of the world also owe their phenomenality to technical factors. It is indeed one of phenomenality's unnoticed foundations that it is technically conditioned. It's not just about considering that everything is technically influenced, as if the technologies had, from the outside, a simple impact on phenomena—which is not wrong. Pierre Lévy also stressed the inanity of the metaphor of impact, which compares technology to a "projectile" striking culture or society.[20] It is about showing that the fact of appearing as a thing is a process of phenomeno-technology in itself and that technology determines, from the inside, the phenomenality of phenomena.

By "phenomenal phenomenality," we mean the way in which being (*ontos*) appears to us (*phainomenon*), inasmuch as phenomenality induces a particular quality of feeling-in-the-world. We call it *ontophany*, in the etymological sense of the term as Mircea Éliade uses it, meaning something that manifests itself to us.[21] To assume, therefore, that any ontophany of the world is a technical ontophany,[22] or at least has a technical dimension, amounts to postulating that there are a priori conditions of perception that are not transcendental as in Kant but technological as in Bachelard. From this perspective, the technological can be defined as an ontophanic matrix, a general structure of perception that a priori conditions the way in which beings appear. As such, this structure does not belong to the organization of our faculty of knowing (it is not an a priori structure of the subject of knowledge), but to the external organization of our technological culture (we propose calling it a techno-transcendental structure). And the technological culture in which we live relies on a technological system.

Indeed, from one technological system to another, it is not only the object of perception that changes according to the groundbreaking materials that are used (for example, wood, steel, gas, electricity, information) or to the invented devices (crank-rod, steam engine, particle accelerator, computer). What changes is the act itself of perception in its phenomenological dynamics, because it is the very phenomenality of beings (that is, their ontophany) that is redefined and renegotiated by technological culture. For a phenomenon of the world, whatever it is, the mere fact of appearing is very different if it appears at the time of wooden machines and water (premechanical ontophany or "eotechnical"[23]); of blast furnaces and steam engines (mechanized ontophany); or of interfaces and networked computers (digital ontophany). Any experience of the world depends on technological ontophany because in each case, the mere fact of appearing—that is, of pure manifestation, or ontophany—consists in offering up to the subject according to perceptual characteristics that are fully conditioned technologically and that make possible experiences-of-the-world at any given time unique and singular, in the sense that this experience is technologically aware of the times. Let's take a few examples.

In the Renaissance, a spinning wheel set up outdoors operates with the quiet and regular squeak of the wooden gears to the easy beat of the spinner's foot or hand, in a calm sonic atmosphere less than or equal to the sound of the wind blowing in the surrounding trees; the body's contact with the wool and the wood—living materials—provides a sense of carnal continuity with nature. The mill wheel turns with the natural movement of the water, whose lapping one hears, while the hydraulic saw mechanically follows its ripples, with occasional sharp screeching. On the horizon, the visual landscape is hardly transformed by all this technical movement and the setups involved. We are in the Renaissance, and we are having an experience of the world whose phenomenal quality is conditioned by a premechanical technology system that is hardly intrusive. This is eotechnical ontophany.

Three centuries later at a coal mine, the force of steam lifts huge metal pistons that are immediately lowered and strike the ground with a deafening din; the boiler releases a moist and significant amount of heat while the sky darkens with the black smoke of burning coal whose odor spreads over several hundred meters; once started, this machine imposes its brutal and systematic cadence, as with locomotives whose characteristic whistle one hears in the distance; prolonged body contact with the cold and inert metal material causes a sense of disturbing strangeness and creates long hours of depersonalizing work; on the horizon, the silhouettes of pistons and long chimneys cut up the sky and draw a new landscape. The phenomenal quality—that is, its perceptual quality—is an experience-of-the-world that is quite different from that of the Renaissance. It is conditioned by the first industrial technological system. That is mechanized ontophany.

Two centuries later, on an artificial screen that diffuses a significant amount of light, a microcomputer displays windows, icons, and menus offered up for interaction. Stimulated by the image, our eyes remain fixed for hours on this silent object that is placed on a table and needs to be inside a built space. From time to time, pretty stereotyped sounds emanate from the machine to alert us to a message or an event. Our hands and eyes are constantly in demand, and thanks to numerous interactions with the system, an immense amount of information can be processed in a single day, making laborious tasks always faster and more efficient, and idle and fun activities always more spectacular and attractive. It's difficult to tell what the weather it is outdoors or what is happening in the vicinity since the object invites immersion. Information can, however, be searched on the Web, or using a cell phone at hand, itself consisting of a screen that is smaller in size. Contact with the landscape no longer exists directly except as digital images offered as wallpaper. Networked interactions, however, enable us to remain in permanent contact with our correspondents, whose messages flood Twitter and Facebook. That's what the possible experience-of-the-world looks like in the digital era, when conditions for exercising perception are conditioned by the digital technological system. This is digital ontophany.

These three scenarios deserve more development, but they are already enough to illustrate how much the technological system of an era shapes, in the manner of a techno-transcendental structure, the phenomenal quality of the world we experience. It is not so much the object of perception that is different but the act of perception itself that has changed since, when having an experience of the world whose ontophanic quality differs, the very means of feeling-in-the-world is overhauled. It's as if it were not the same world, as a world, that we were experiencing because its process of its phenomenal manifestation is inseparable from the ontophanic qualities, of which technological devices are made, which compose and condition it.

If ontophany is the manifestation of being and if it depends on the surrounding technological culture, then the quality of ontological phenomena of the world is always conditioned by a technical reference system. And so every change in technological systems needs to be interpreted as a change of ontophany where the methods of perception are redefined. The possible being-in-the-world emerging from eotechnical ontophany and characterized by the silence of the instruments and the bodily proximity of nature is not the same as being-in-the-world that emerges from mechanized ontophany and is characterized by the violence of the machines and the widespread mechanization of bodily existence, or the same as that being-in-the-world emerging from digital ontophany that is determined by the speed of computation, the fluidity with which procedures are executed, and the immersion in interfaces.

All of this shows to what extent experience is in itself a phenomeno-technological construction. As a perceptual interaction between subject and phenomenon, it is always technologically produced—and not simply technologically mediated[24] or influenced. By "phenomeno-technical," we must thus understand that a technological principle can condition the phenomenality of a phenomenon. Phenomeno-technology is the technological construction of ontophany. This is why the technological is presented as an ontophanic matrix, namely, as a phenomenological mold, produced by culture and history, in which our experience-of-the-world can take place.

Such a conception can be related to the aesthetic constructivism of Anne Cauquelin as it emerges in *L'Invention du paysage*, which I read even before Bachelard and is the initial source of inspiration for the hypotheses developed here. There is nothing natural about landscape; it was invented in the Renaissance at the same time as the laws of perspective:

> The question of painting stands there: it projects before us a 'surface,' a form into which perception flows. We see perspective; we see paintings; and we do not see, nor can we see, otherwise than according to the artificial rules set up at this precise moment, the moment when, with perspective, the question of the painting and the landscape emerge.[25]

A form where perception flows—that's what the ontophanic matrix is. The technological concatenation of perspective is the ontophanic matrix of nature; it conditions the phenomenality of nature, that is, the way nature appears to us (at least visually). Oscar Wilde had noticed this well before anyone else, and not without impact:

> For what is Nature? Nature is no great Mother who has borne us. She is our creation. It is our brain that she quickens to life. Things are because we see them, and what we see, and how we see it, depends on the Arts that have influenced us. To look at a thing is very different from seeing a thing. One does not see anything until one sees its beauty. Then, and then only, does it come into existence. At present, people see fogs, not because there are fogs, but because poets and painters have taught them the mysterious beauty of such effects. There may have been fogs for centuries in London. I dare say there were. But no one saw them, and so we do not know anything about them. They did not exist till Art had invented them.[26]

If art invents fogs and brings them into existence, it is good that culture—artistic culture, in this case—has the power to generate phenomenality. And what happens with technology is at least as true as what happens with painting, or much more, for not only does what we see depend on the technologies that have influenced us but, let's repeat it, the technologies that we live with at a given time condition the very modalities of the possible manifestation of the world at that time. To observe the sky in the age of the steam engine is not to experience the sky ontophanically the way one does in the age of digital interfaces, even if it is (perhaps) the very same sky and even if the sky is not an artifact. Each technological system creates different ontophanic conditions, that is, different material conditions of phenomenal manifestation, which are particular and particular to it (as the technological system of a given period) and form a particular "perceptual surrounding" (in the sense of Jakob von Uexküll's *Umwelt* or "self-centered world"). To each technological system, there corresponds an ontophanic grid of the real. The history of technological revolutions is the story of ontophanic revolutions because there doesn't exist any phenomenon on earth outside of technological conditions in which phenomena are possible. Our perception of reality is the result of what our mind builds from the technological operators of reality that it has at its disposal at any given time in history. Technology is thus indeed a form where perception flows, a techno-transcendental structure that produces the conditions of reality's phenomenality. The world is the result of phenomeno-technological casting.

Viewed from this angle, the real is never universal and substantial. There is no ontology of the invariant, as if *being* had always been there, to the point where, as Heidegger sees it, it could have fallen into oblivion. There are no invariant ontological structures, and there never were any. There are only, and always will be, changing phenomenological structures—what Pierre Lévy calls "transcendental history."[27] Ontology can only be historical and cultural phenomenology or, as Peter Sloterdijk suggests, an onto-anthropology because the real or a *being* is always particular and accidental, sensitive to the technological conditions of the times. Being-in-the-world or being-there (*Dasein*), it is not a general metaphysical condition detached from the conditions of the century. Being-in-the-world or being-there is simply not the same thing if one lives in the premechanical, technological system or the digital technological system. The stuff of the world is the same age as our technologies.

Dialectics of the Apparatus and of Appearance

What is so remarkable in the sensitivity of an age is the concomitance of ideas. In all eras and in all domains, new ideas flourish always almost at the same time in the mind of a handful of men and women whose destinies do not always meet. Alexander Graham Bell is, as we know, the inventor of the telephone, but when we look more closely at the story, we realize that inventors Antonio Meucci and Elisha Gray are no less so.[28] Similarly, John von Neumann is known as the inventor of the theoretical computer model, but historians know John P. Eckert and John W. Mauchly are no less. The list of the adventures of concomitant ideas is long.

The one that interests us here is philosophical. It happens in 1931 as two works are published. On one hand, in France, Gaston Bachelard publishes his article "Noumenon and Microphysics" in the journal *Recherches philosophiques*, in which he introduces the concept of *phenomeno-technology*; on the other hand, Walter Benjamin publishes his celebrated article, "A Short History of Photography," in three parts in Germany in *Die Literarische Welt*.[29] A few years later, in an even more striking coincidence, these two authors reinforce and develop their respective intuitions, each through a landmark work: the first in *The New Scientific Spirit*, published in 1934, and the second in "The Work of Art in the Age of Technological Reproducibility," the first version of which was written of 1935.[30] What is striking is the concomitant emergence of the concept of phenomeno-technology, whose enunciation takes us back to Bachelard, but that each of these two authors in his own way helps to clarify: the former by analyzing the work of modern science using nuclear physics, the latter using photography to investigate the nature of modern art. In what ways does Benjamin's thesis about the work of art meet Bachelard's phenomeno-technology, and how can it help us formulate the general principles of a techno-transcendental phenomenology?

A first answer is very short: just as Bachelard introduces technology at the heart of scientific activity, Benjamin introduces technology at the heart of artistic practice. For Benjamin, indeed, it is time to question "in all its ponderous vulgarity ... the philistine notion of art, dismissive of every technical consideration," this "fetishistic and fundamentally antitechnological notion" that with the invention of the technologies of photography "sensing its doom."³¹ In analyzing what he calls "the expansion and decline of the photography,"³² at least during the aesthetic decline we can assume photography was in (very temporarily) the 1920s because of the excessive expansion of the photographer's occupation at the expense of the painter's, Benjamin shows that we have lost the beauty of old photographs, those of early daguerreotypes, those "first photographs … of unapproachable beauty."³³ According to him, these images had the merit of making beings appear with a technological phenomenality that was in tune with the natural phenomenality of the world in the sense that on these shots, the characters had an "aura" of the same sort that things have, or the phenomena of the world, in their natural states:

> What, then, is the aura? A strange tissue of space and time: the unique appearance of a distance, however near it may be. To follow with the eye while resting on a summer afternoon a mountain range on the horizon or a branch that casts its shadow on the beholder is to breathe the aura of those mountains, of that branch.³⁴

Such a notion cannot leave us indifferent. The aura is the uniqueness of an apparition as a moment of phenomenal truth. The aura is ontophany's other name. In other words, not only does the world's natural phenomenality offer itself up in an aura, but also early photographs are able to restore it: "In the fleeting expression of a human face in the old photographs, the aura beckons from early photographs for the last time. That is what gives them their melancholy and incomparable beauty."³⁵

Early twentieth-century photography, however, lost the ability to restore the aura of things. Becoming an essentially social technology of mass reproduction ("photography took over from painting"³⁶), it endlessly spews tasteless images in order to meet the new need to fill family albums (for this, we use a host of ridiculous props seeking to imitate painting: columns, carpets, draperies, palms, tapestries, and so forth). This means we can talk about photography's aesthetic decline just when it is experiencing its first great social expansion. In Benjamin's eyes, clearly this decline is the aura's decline. Photography, at least when it is solely a function of "technological reproducibility," depreciates the here-and-now, that is, what constitutes the ontophanic uniqueness of the aura, as phenomenal authenticity.

What happens, then, with the decline of the aura is that we are witnessing "changes in the medium of present-day perception"[37] because perception is technologically determined by the apparatus: "The decisive thing about photography is the photographer's relationship to his technique";[38] the photographer's eye is above all technical expertise or the scientific knowledge of the machine that captures light. This expertise is essential in order to make the aura emerge. And in terms in which something like a phenomeno-techique of art surfaces, Benjamin writes:

> It is indeed a different nature that speaks to the camera from the one which addresses the eye; different above all in the sense that instead of a space worked through by a human consciousness there appears one which is affected unconsciously. It is possible, for example, however roughly, to describe the way somebody walks, but it is impossible to say anything about that fraction of a second when a person starts to walk. Photography with its various aids (lenses, enlargements) can reveal this moment. Photography makes aware for the first time the optical unconscious, just as psychoanalysis discloses the instinctual unconscious.[39]

Just like a physicist's instruments can make visible the microphysical world, the camera's technological provisions make it possible to restore the aura of things: "[Technological reproduction] can use certain processes, such as enlargement or slow motion, to record images which escape natural optics altogether."[40] Equipment that holds the power in this way to make things appear must therefore be called, echoing Bachelard, a phenomeno-technological apparatus. Such is the camera, or even the film camera,[41] but not only; particle accelerators, telephones, and computers are other typical examples, which we will come back to.

Walter Benjamin's artistic aura thus meets Gaston Bachelard's scientific phenomeno-technology. It allows one to support the hypothesis of a general phenomeno-technology even more solidly because to affirm that the world is the fruit of phenomenological casting or that technology is an ontophanic grid conditioning the phenomenality of reality is not simply to say with Bachelard that scientific technologies produce physical phenomenality, or with Benjamin that artistic technologies produce sensory phenomenality. It means that all the technical/technological procedures of an era, as defined and unified in a technological system that is historically determined, produce the singular general phenomenality that creates the world of that time in its ontophanic dimension. That means we are to the world only as given to ourselves in an ambient phenomeno-technology having the precise value of a *perceptual environment* or *Umwelt*. Phenomeno-technology is more than just a matter of technological equipment; it is a fact of reality, in the sense that it is immanent in the stuff of the world—provided that the stuff of the world changes with history.[42]

It then becomes possible to reformulate Walter Benjamin's theses in terms of a general philosophy of technology defined as techno-transcendental phenomenology. From this perspective, the decline of the aura dear to the author of the "Short History of Photography" is merely the visible sign of the current ontophanic revolution, the one related to the technological system of mechanization that includes the camera as one of its many inventive incarnations. This decline is not so much an aesthetic event as it is an ontophanic one. Everyone is free to judge its aesthetic value. But what matters is that through the idea of a decline of the aura, Benjamin does nothing more than grasp, in the spirit of the times, the effect of change in the ontophanic matrix that is characteristic of the movement of history.

That this change operates temporarily, to the detriment of certain aesthetic values, alters nothing. The central event, which affects the world in a systemic way, is the ontophanic revolution, that is, the change in the phenomenal status of the world. The resulting disruption of aesthetic values is only one side effect among others. In the world of art, the alleged decline of photography is only one of the many aspects of the ontophanic revolution triggered in society as a whole by the mechanical technological system.

Using the example of the camera, one can see better the intimate relationship woven into the culture of an era between a technological system and the ontophanic matrix it generates. Walter Benjamin could not make out that relationship, which only a techno-transcendental phenomenology can reveal, even if he intuited it. On the other hand, the work of French philosopher of art Pierre-Damien Huyghe on the technological conditioning of art, in constant reference to the author of the "Short History of Photography," helps to understand better what Benjamin could only glimpse and comes close to this techno-transcendental phenomenology. Indeed, although his discourse is never dissociated from aesthetic concerns[43] (which rarely allows him to approach technology itself), Pierre-Damien Huyghe endows the concept of apparatus, directly derived from Benjamin's thesis, an increasingly important role as his work grows.

In 2002 in *Du commun* in a chapter on the question of art, evoking "the technological fundament of being,"[44] Huyghe writes, "Technology is made to make something possible; it constitutes the overall domain of the elaboration of culture."[45] One could read in that a simple Simondonian reminder, but as we will see, he is sketching out more than a phenomeno-technological intuition. On the model of photography, the *apparatus* of the camera is still conceived of as a device that captures scenes;[46] and the central thesis, in line with Benjamin, is that of art's conditioning by technology: "Thus we will say that doing art is to take risks with a technology, a *savoir-faire*. ... The art is to pursue or to push a technology (drawing, for example) outside its area of effectiveness.[47]

Nevertheless, a more general hypothesis seems to emerge:

> That this be so—that the apparatus should impose itself on the thought of beings in the world—is not just a fact of art in the restricted sense of the term. The equipment is an essential pipeline of existence.[48] From this vantage point, there is no sense in opposing a pure art of devices (drawing, painting) to an impure technology (photography, cinema).[49]

In 2003, a short, inspired text about "forms of perception an apparatus engages"[50] enriched Pierre-Damien Huyghe's research. On this occasion, he offers a more general definition of the concept of apparatus: an apparatus is not a technological object like any other, but "a technological modality separate from the tool and the machine."[51] The fetishized example remains that of the camera. Although *serving* to produce images just as a device would ("economic logic"), it would also be able to produce perception and enter into the phenomenological regime of the apparatus ("aesthetic logic"). But now with the "aesthetic logic," the question of art returns, a question Huyghe does not care to leave behind, as if it were impossible for him to consider technology apart from an aesthetic use and, hence, in a contemplative way of thinking. The ability of the apparatus to produce perception, and therefore phenomenality, is nonetheless reaffirmed: "The particular characteristic of the apparatus" means "that it has potential perception, a particular form of sensitivity,"[52] which can be referred to as a "capacity of the apparatus to fashion sensitivity," "the power to create some world, to make worlds."[53] Here we are not far from the idea of a general phenomeno-technology, which this statement confirms in its way: "Whenever a new apparatus appears in human history, it's a way of splitting up sensory frameworks which risk being destabilized."[54] One could not say it better, except that *rebuilt* would be better than *destabilized* in order to avoid any negative value judgment. But the author quickly returns to a contemplative phenomeno-technological art:

> Human beings are beings of the sort that the forms from which their experiences are structured in a shareable world—a common world—are not only internal but also external, historical and technological. This is why we are sensitive to various spatio-temporalities, that of pictorial perspective for example, but also that of photography or even cinema. ... What matters is to understand that, for the most part, we see less what we want than what lies in the possibilities of these devices.[55]

In 2004 in *Le Différend esthétique,* Pierre-Damien Huyghe goes one step further by more generally analyzing the intimate link between existence and technology:

> To say that art constitutes the schemas that a technology needs to free itself of its uses and join itself to a thought is to say, in the context of this analysis, that with each new technological shift or new reinstrumentation of the world, what disappears in the way old art schemas and old expressions of experience is always linked to some form of space and time, that is to say, to an encompassing form of sensitivity. This statement may seem Kantian: it is not quite, precisely because of what it implies about the historicity of the forms in question, the possibility of these forms' work. Here art is about making an apparatus of time and space work noticeably, or, if you prefer, the concealed spatio-temporality of the apparatuses in use. Because that's where a technological shift can be considered decisive for an era: it modifies the nature of being in the world (it touches on the *being here* of being) because it touches this being's spatio-temporality (upon the form of its *being*).[56]

That a *technological shift* can change the nature of being in the world makes the case for the hypothesis according to which technologies are ontophanic matrices. By technological *shift*, one must understand technological revolution, in the sense that we defined and analyzed it in the mechanical revolution or the digital revolution. The relationship between a technological system (an era) and the ontophanic modality it engenders (a way of being in the world) is then established. Starting from a phenomeno-technology of art inspired by Benjamin, Pierre-Damien Huyghe comes close, by way of aesthetics, to a general phenomeno-technology inspired by Bachelard's pathways of epistemology.

After 2005, this point becomes even clearer. In a "time of widespread instrumentation of perception,"[57] Huyghe writes, "[social reality] takes shape thanks to the presence and dynamics of a certain number of technologies and ways these technologies manifest."[58] This way of taking shape is "appearance, that is to say technical phenomenality."[59] And so we can see how the concept of apparatus is the basis of Pierre-Damien Huyghe's phenomeno-technology: "An apparatus—I always start with Benjamin to advance this proposal—is a device whose operation can create consciousness."[60] The philosophy of technology that emerges from the aesthetic thought of Pierre-Damien Huyghe is thus a philosophy of the apparatus. He phrases it most generally in *Modernes sans modernité*: "Technology has a complex relationship with 'appearance.'"[61]

It is, therefore, of course, on the dialectic of the apparatus and appearance, in the sense that an apparatus shapes the basis of the world's ability to offer itself up to perception, that we can develop our hypothesis of a general phenomeno-technology. This hypothesis is the point of departure for our philosophy of technology conceived of as a techno-transcendental phenomenology, whose consequences cannot all be drawn in the frame of this book, but which is alone in being able to grasp this *essence of technology* that the haters of technology of the twentieth century missed.

It rests on the idea that all technological objects, although some more than others, are apparatuses, that is, phenomeno-technological apparatuses. There is no need to distinguish between these two systems: one (impure) would be that of the instrument (a tool or machine not operating fully and restricted to the economic logic of use); the other (noble), that of the apparatus (a phenomeno-technological device making use of all the object's possibilities and rising to the aesthetic logic of full use).[62] Any technological use is perforce a phenomeno-technological use, even if its phenomeno-technological level is weak or invisible. For the artist and the architect, a brush and a pencil, a Rotring pen and a compass are no less devices than a daguerreotype or a graphics tablet, AutoCAD software, or Instagram. For the writer, an Egyptian *calam* or a fountain pen is no less a device than a stylograph or a typewriter in the mechanical era, or a computer, a word processor, and touch pad in the digital age.

All our objects are part of the world and participate to varying degrees phenomeno-technologically in the ontophanic process of the real—and not only cameras that capture a scene. Spinning wheels or steam engines, hydraulic saws or blast furnaces, automobiles or telephones, computers or the internet: all of these technical and technological apparatuses are, to varying degrees, machines that make the world appear and change the nature of the experience that we can make of being. The era of technological apparatuses is not just modernity. The age of apparatuses is as old as humanity. We have lived forever in an augmented reality.

The Model of Telephonic Ontophany

The example of the telephone is edifying. It deserves a "A Short History of Telephony" à la Walter Benjamin to demonstrate how, at the end of nineteenth century, the sudden eruption of the human voice from a box of electrified wood upset the phenomenality of the world and the phenomenality of our relations to others, as well as societal relationships—crystallizing the idea of an ontophanic revolution. For this, we must return to the first uses of the telephone.

Intended to compensate for the deaf and the hearing impaired, Graham Bell's telephone is invented on February 14, 1876, at 2:00 p.m. (patent filing date). On March 10, 1876, the first sentence relayed by electric telephone was transmitted between two rooms of the top floor of a house in Boston, using these famous words that Bell addresses to his assistant in the other room: "Mr. Watson, come here—I want to see you!" The telephone is born, and this episode is considered its historical origin, despite the many quarrels about its genesis. The Bell Telephone Company (AT&T's ancestor) was established in 1877, and by the end of August of that year, thirteen hundred telephone sets are in use in the United States. The same year, Thomas Edison invents the microphone, which makes it possible to appreciably improve the apparatus.

Presented at the Académie des Sciences in France in October 1877, the invention spreads rapidly through Europe. At first, as Robert Vignola points out, nobody knows what purpose the telephone might serve[63] (much like the social media Twitter in its early years). There are those who believe in it and those who see it as a scientific curiosity without use, if only because the voice is not always audible. At first, the telephone is considered a simple "machine to deliver information over distance rather than to transmit conversations."[64] No one can imagine using it for work: "As an object of leisure for a well-off class, the telephone is considered one of its frivolous privileges."[65] When it is first used, however, what strikes people—and this will not be surprising—is precisely an ontophanic innovation: "To be heard without being seen, now that's an exciting novelty: the situation mostly inspires comedians of that time, as a prolific source of burlesque and libertine *quid-pro-quos*."[66]

What does this mean in philosophical terms? Precisely that this new object upsets the very phenomenality of our relationship to others (the ontophany of others). Never before in history had it been possible to hear the sound of the human voice without simultaneously seeing the face of a human before one's eyes. Ontophanic vertigo. Phenomenological revolution. Such a perceptual experience simply never had been possible before. That others can give themselves up to me as a concrete auditory presence while being just as concretely visually absent is radically novel for my senses and my awareness, a novelty for which there is no perceptual culture of reference. This new ontophany of the other—unheard of (one dare say)—is made possible by a simple technological object, an electrified wooden box, a device, an apparatus. Here at work is the phenomeno-technological process in all of its power as a quasi-demiurgic process, that is, which creates a world. A telephone is also a form in which perception is cast, that is, a techno-transcendental structure.

From a philosophical point of view, the invention of the telephone should therefore be considered one of those many ontophanic innovations that accompany the development of a new technological system—here, the mechanical technological system—that is, one of these disruptions of our potential experience of the world. As such, in its early years, it could not fail to produce a socioperceptual shock, a phenomenological rupture in ontophanic culture. The exceptional testimony of Pauline de Broglie, countess de Pange and sister of the physicist Louis de Broglie, is highly illustrative in this respect. In an autobiographical narrative, *Comment j'ai vu 1900*, published in the 1960s, she recounts her childhood memories and, among other things, the telephone setup in her parents' mansion in Paris around 1896 to 1898, when she was barely ten years. What she reports perfectly translates the ontophanic disruption the telephone caused in the daily lives of its first users:

> The telephone apparatus was put at our place in a pass-through lounge. It was made of rosewood and was nailed to the wall. It sort of looked like a small toilet paper box in a w.c. Two earphones hung on hooks on each of its sides, and in the middle a button you pressed to get through to the central station. And this more and more furiously because the answer was slow to come. We spoke in front of a small board that my mother wiped down carefully "to remove the miasmas," she

said after each conversation. The ring was shattering and could be heard throughout the house. But we did not run to the telephone! A servant whose task it was picked up the earpiece, inquired what was wanted, and went looking for the asked-for person. I could hear the ring from my room, and the strange call, such an exotic a sound: *Ello! Ello!* my mother strove to pronounce in English. Naturally there were no telephone books since there were no telephone numbers. The request was direct, and there were continuous battles with the "telephone *demoiselles*." After half an hour agitating and discussing, my mother was in hysterics and got migraines, but she always went back, whereas my grandmother never even wanted to go near the device. She hated this way of *talking to others without seeing* them. I should add that, well after 1900, until I was twenty, I was not myself allowed to pick up the earpiece! A well-behaved young lady was not to answer the phone until one had made sure of the caller's identity. A well educated young man would never have allowed himself to call a young lady on the phone without going through her parents.[67]

This excerpt is very valuable in many ways. We note this simple and spontaneous wording, whose precision is remarkable: *talking to others without seeing them*. There in just a few words is exactly what, from a perceptual point of view, is the new ontophanic modality that the telephone introduces into relationships with others—to the de Broglie grandmother's chagrin. But whether one likes or doesn't like this way of being-with-others, it is a new way of being-in-the-world. And it is going to change everything because by becoming of importance to industry, the telephone is rapidly gaining ground in the world, entering into professional use and modifying social practices on a large scale. In the United States, where its development grows fastest, Herbert N. Casson is already speaking of "telephonization of life" in 1910:

> What we might call the telephonization of city life, for lack of a simpler word, has remarkably altered our manner of living from what it was in the days of Abraham Lincoln. It has enabled us to be more social and cooperative. It has literally abolished the isolation of separate families and has made of us members of one great family. It has become so truly an organ of the social body that by telephone we now enter into contracts, give evidence, try lawsuits, make speeches, propose marriage, confer degrees, appeal to voters, and do almost anything else that is a matter of speech.[68]

One might be tempted to apply exactly the same words to the societal disruptions that we are experiencing today in the digital era.

What circulates in this huge, cabled network of telephone lines that is put in place between the two world wars is nothing less than this new ontophanic culture of *talking to others without seeing them*. As soon as the number of subscribers to the central exchanges grows, this new culture spreads and progressively merges with the general ontophany to become, in its turn, commonplace and ordinary. Today, whether one enjoys talking on the phone more or less, there is no longer any de Broglie grandmother who will not go near the device. Living without a phone—"this good old phone that we have so well integrated into our lives that it seems natural to talk to each other without seeing each other,"[69] as Serge Tisseron stresses—seems even to have become impossible. Along with an old technology, it is an old way of being in the world that has disappeared. The new telephonic ontophany has swept away the previous ontophany, and no one sees this as an issue any longer. Being-in-the-world telephonically is part of the world of the dominant ontophanic culture of the twentieth century. Telephonic ontophany became a *naturalized* perceptual culture.

One hundred years later, at a time when the computer and the internet take their turn insinuating themselves into the entire social body, the same process starts all over. And the same de Broglie grandmother now flees screens and interfaces and worries about their impact. This new world, however, will settle in, and it will establish a new ontophanic revolution. After having learned to talk on the phone to each other without seeing each other, today we are learning to bond without talking or seeing each other, as we do on Twitter and Facebook, and it does not mean we are "alone together."[70]

Each generation relearns the world and renegotiates its relation to reality by means of the technological devices at its disposal in its own sociocultural context. That's why the *generational digital divide* is perhaps only a poor phenomenological interpretation. If the oldest people sometimes have trouble adopting new technologies (which seems less and less true), it is because their relation to reality is simply cast in an ontophanic matrix other than the digital one. And if the youngest (the so-called digital natives) are generally more comfortable with interfaces (Michel Serres's Thumbelinas[71]), it is simply because they do yet have the perceptual structures (they are phenomeno-technological virgins) and because devices around them for them are the only way to acquire perceptual structures and, therefore, to come into the world. Being a digital native is to be made in the same phenomeno-technological mold as the interfaces that connect them to reality. Being a digital native means having acquired the faculty of seeing the world appear by being digitally outfitted. Being a digital native is, properly, to be born of and through the digital because coming into the world is not enough to be born in the world. Only the technological objects that surround us enable us to be born into the world in a phenomenological sense. One also learns to exist with objects—in the narrow sense, where *to exist* here means *to be made present in the world*. To be, then, is to be born with technology. To be is to be *tech-born*.

There is no generational digital divide. There are only dated ontophanic matrices that overlap and coexist. This is how we confirm that perception is the least natural thing in the world; in each era, the act of perceiving is learned through existing technologies. This is why older people are most often nostalgic for the place of old objects or that everyone keeps some childhood objects for life because objects that come from another era have an invaluable phenomenological value. They bear the imprint of another phenomeno-technological time, a time when the world did not have the same "aura," a time when feeling-in-the-world was not exactly the same ontophanic flavor because the vigorous perceptions of that time, inflected by the pleasure of first times, were cast in a world of sensing of technologically dated objects, but in which flowed the still new pleasure experienced by simply being present in the world.

4 The Life and Death of the Virtual

> What does the virtual alter in the very act of "seeing"?
> —Philippe Quéau, *Le Virtuel: vertus et vertiges*

Like all previous revolutions, the digital revolution is an ontophanic revolution. Of all those that have occurred in the course of history, it is one of the most penetrating and spectacular. Certainly technologies have always conditioned the phenomenality of the world—this is the meaning of general phenomeno-technology argued here—but this has never been as true as in the era of digital technologies. On this point, no technology had ever transformed the way in which beings and things appear to us as phenomena. In the past thirty years, the perceptual shock caused by the digital is so great that without exaggerating, one can speak of the real phenomenological trauma—in the neutral sense (and almost clinical) of the term—that took place in our experience-of-the-world. The network interfaces of the digital technological system, those of our computers, consoles, smartphones, tablets, and more generally of all our networked things, are the new ontophanic devices of our time, that is, the new phenomeno-technological apparatuses through which the world appears to us today. They established themselves in just a few decades, and like the machines of the mechanical era earlier, they have profoundly changed our way of being-in-the-world, establishing a new mold into which our perception flows and creating a new environment of perception or "own world" (*Umwelt*).

An abridged version of this chapter, "Contre le virtuel: Une déconstruction," was published in *MEI: Médiation Et Information*, no. 37 (2014): 177–188.

This started with the appearance of microcomputers in the mid-1970s, which enabled us to become *computerized*; it continued with graphical interfaces of the 1980s, which made screens *worlds of images*; it spread with the rise of *cyberspace* in the 1990s and then the triumph of Web 2.0 and mobile devices after 2000, which ushered us into the *global village* and the ubiquity of digital use. Today, after several decades of learning about and immersion in digital interfaces, we can say that we have changed worlds—not in the sociological sense of social structures (which is also true), but in a philosophical sense inasmuch as we changed perceptual structures (in the sense of techno-transcendental structures). The contemporary world of the twenty-first century emerges from a phenomeno-technological casting of the digital sort. We are, hereafter, present to things and beings only as long as they appear to us through and around digital devices. To understand this as yet unknown, novel, digital ontophany, then, for a philosopher of technology, it amounts to interrogating digital phenomenality itself. Of what phenomenality are digital phenomena capable? How do digital beings manifest? What does their being consist of?

In order to answer this, it will be necessary first to examine the hypothesis of the virtual and the metaphysical imaginary that the hypothesis carries. On the way, one has to listen to the attempts the first thinkers about *cyberspace* made to characterize digital ontophany. How an old word from medieval metaphysics such as *virtual*, fated through a long theoretical tradition to cast more confusion than light, has all on its own come to characterize and summarize the phenomenology of a new world—of informatics, computers, and the internet. This never ceases to be surprising and deserves to be deconstructed. That is the purpose of this chapter, which means to show that the relationship between technology and reality has never been as hot as with digital technology.

The Genealogy of the Virtual: Philosophy, Optics, Computing, Psychoanalysis

It can't be repeated enough: the term *virtual* was not coined in the computer age. It is a word from the language of philosophy whose long history does not always make it easier to use it rigorously.[1] Translated from medieval Latin *virtualis*, the term is used for the first time in the Middle Ages in order to translate the concept of *potentiality* (*dunamis*) of Aristotelian scholastic philosophy as against *actuality (energeia)*. In Aristotle, potentiality and actuality are two modes of existence: either a thing exists *in actuality* or it exists *in potentiality*. When it exists in actuality, it is effective and productive. When it exists in potentiality it is only in a potential state; it may produce or be realized but has not actually been achieved. In *The Metaphysics*, Aristotle defines these two states as follows:

> What we mean can be plainly seen in the particular cases by induction; we need not seek a definition for every term, but must comprehend the analogy: that as that which is actually building is to that which is capable of building, so is that which is awake to that which is asleep; and that which is seeing to that which has the eyes shut, but has the power of sight; and that which is differentiated out of matter to the matter; and the finished article to the raw material. Let actuality be defined by one member of this antithesis, and the potential by the other.[2]

Let's say I want to carve a wooden statue of the god Hermes. As long as the statue is not realized, "Hermes exists in potentiality in the wood," but as soon as I start to carve, the sculpture of Hermes exists in actuality in the wood. "'Actuality' means the presence of the thing not in the sense which we mean by 'potentially.'"[3] In this sense, all of our faculties (for example, seeing, feeling, thinking) are potentialities whose essential feature is that they can at any moment be actualized. When I close my eyes, sight exists in me in potentiality (that is, virtually), whereas when I open my eyes, it exists in actuality (that is, currently). It is this state of potentiality (to be actualized) that medieval philosophers translated as *virtualis*, from the Latin *virtus*, "potency, energy, merit, virtue." As Gilles-Gaston Granger accurately points out, "We see that nonactuality as introduced by Aristotle is in no way the opposite of reality, even though it is actualization that constitutes perfection and achievement of all kinds."[4] Thus, "until the seventeenth century, virtual designates potential as opposed to actualized,"[5] and did well into the classical age, since in 1926 it was still André Lalande's dictionary definition: "Virtual is what exists only in potentiality and not in actuality."[6]

So now let's clarify this misunderstanding:

> *Virtus* is not an illusion or a fantasy, or a mere eventuality, thrown to the limbo of the possible. It is very real and in actuality. *Virtus* fundamentally acts. ... The virtual is neither unreal nor potential: the virtual is of the order of reality.[7]

Indeed, in the philosophical meaning that has been Aristotle's for centuries, the virtual is none other than an ontological regime, a particular way of being real, which amounts to existing without being manifest. So there is absolutely nothing in the philosophical concept of the virtual that legitimates confusion with the nonreal, as occurs so often. "Unlike potentiality, which is a maybe, in the future, the virtual is present, in a real and actual way, though hidden, underground, inevitable."[8] When a child plays hide-and-go-seek in the garden, her presence does not become unreal; she is actually in the garden but in a virtual state, that is, not manifest. Similarly in a sports competition, when an athlete runs in the lead but has not yet crossed the finish line, we say of her that she is, at this precise moment, "virtually a gold medal." That means that her dominance is very real, though not yet fulfilled, not yet fully manifested or phenomenalized (and may not be in the end).

Nevertheless, in parallel with this initial philosophical use, there developed in modern times a scientific use of he term *virtual* in the field of optics, the part of physics that deals with light and vision. Rather unknown to the general public, this use is at the origin of the arbitrary and erroneous distinction between virtual and real that we suffer from so much today. What is it about? For a physicist, an image is by definition an intangible reality, that is, a signal detected by the eye. In this sense, to speak of a *photographic image* is a mistake, since a photograph is a palpable object and not an image (unless one considers it a materialized image). In optics, therefore, an image is either a *real image*, that is, an image that can be seen and collected on a screen (the image, for example, of a luminous object that strikes our retina or an image that comes from a slide projected on a wall or even a television image); or a *virtual image*, that is, an image obtained using an optical device (a magnifying glass, binoculars) and perceived by the eye, but not one you can collect on a screen because it exists only in the apparatus that generates it.

In both cases, from an ontological point of view, we are dealing with two sensory realities, each perceptible to the eye. The only difference lies in the physical status of the image: one is the real image of a real object, and the other is an artificial image, such as the sort generated by a magnifying glass. Philosophically speaking—no offense to the physicist—the virtual image is very real. What characterizes it is not a hypothetical absence of reality but only its artificiality, that is, the fact that it is both produced technologically, using a device, and nonexistent without the device that produces it (to the extent that it cannot be pulled onto a screen that is not part of the device).

As such, the optician's virtual is not of the same order as the philosopher's virtual. Whereas the philosopher's virtual is a way of existing without manifesting, the virtual in optics denotes a very manifest way of existing. What it introduces that is new and is not present in the philosopher's sense of virtual is the notion of artificiality or artificial synthesis: the images made using optical instruments are technologically synthesized images. It is this meaning, which is quite distant from the initial philosophical meaning and indeed rather indefensible from a conceptual point of view, that denotes images that are said to be virtual.

The third meaning of the term arises from that. This time it is technological; it evolves during the second half of the twentieth century in the field of computer science, in expressions such as *virtual memory*, *virtual machine*, *virtual server*, and *virtual reality*. In the world of computers, one calls *virtual* any process capable of simulating digital behavior by using programming, regardless of the physical substrate on which it (paradoxically) relies. Thus, we speak of virtual memory for an "idealized abstraction of the storage resources that are actually available on a given machine" address—space that is not restricted to the physical dimensions of storage devices[9]—or of a virtual machine for "an efficient, isolated duplicate of a real computer machine" that has "no direct correspondence to any real hardware."[10]

For example, thanks to Virtual Box software, a virtual machine licensed as open software, one can easily simulate (in this case "emulate") the Windows operating system on the Mac OS X system, as if one were just launching any other program. And since everything can be reduced to computation, everything can be simulated computationally. In this case, a virtual computer is nothing other than an artificial species, in the sense that virtual memory is artificially synthesized memory and that a virtual machine is an artificially reproduced program. The artifice here rests not as in optics on light radiation technologies but rather on computer programming, that is, on algorithms and computer languages. In the country of code, the programmer is king; like a demiurge, he can simulate, synthesize, or recreate anything. He can even create virtual images—images that do not exist outside the computer equipment where they come into being.

The virtual computer is therefore *simulational*, in the technical sense of the word, that is, the resulting programmable manipulation of information—and not to be confused with simulacrum, which would return us to Plato's cave of lies and mirages, reducing the objective technical nature of the virtual to a fantastical metaphysics of illusion. What is simulational is, on the contrary, entirely real; one can find concrete applications in the operational efficacy of a flight simulator, the scientific accuracy of computer-aided design software, the amazing realism of video games. Mostly neglected, in particular by those who like to confuse it with some metaphysics of the unreal, this strictly technological meaning of the word *virtual* is the only objectively acceptable one, and therefore the only we will be retaining.

Nevertheless, to complete this genealogy, we must add a fourth and last meaning of the term *virtual*: the one that has imposed itself in French psychoanalysis for about a decade now. We owe it in large measure to the thinking of Serge Tisseron, a pioneer of the psychological approach to computers in France and to Sherry Turkle in the United States. For Tisseron, the virtual is a dimension of psychic life, which is different from that of the imaginary. It is neither about an ontological regime in philosophers' sense nor a physical or technological process as computer scientists mean it, although it is closer to the former than the latter:

> There exists some psychic virtuality in human beings, which is not the imaginary. The imaginary refers to an object that does not exist, whereas the virtual concerns all of our expectations and our representations that pre-exist any encounter with reality.[11]

Or: "The imagination develops in parallel to the real world and makes no pretense of encountering it. On the contrary, the virtual prepares one to it and is intended to update it."[12] In psychic life, the imaginary belongs to fiction, whereas the virtual belongs to the reality-of-potential-realizables. From this perspective, the virtual designates the part of our imaginary world that can be realized in the real world. For example, when I imagine someone on Facebook whom I have never met in offline reality, I'm producing this person with a mix from my imagination (because I associate desires and fantasies, worries or anxieties with this person), but that does not mean that it's fiction (because that person is real, and I can verify that by interacting online with him or her through my *friends* network). This kind of representation according to Tisseron is about psychic virtuality—the inner world of representations that we attach to a being and who, when a meeting with this being takes place in face-to-face reality, must be psychically reworked to reflect this reality. In other words, psychic virtuality is an imagined anticipation of reality (possibly sustained by digital devices) called to reshape itself when in contact with this reality, in the impulse it involves—wherefrom what we ourselves have called "misunderstanding of the virtual,"[13] according to which what we imagine in online worlds never conforms to what we experience in offline reality. There is always an offset, which results from the confrontation between what we anticipate in our imaginations and what we then experience as real. It is precisely this gap, this interstice, that Serge Tisseron calls *psychically virtual* since it is potentially actualized. We could just as well call it a *fantasy space* or *fantasy world*, taken in a very broad sense to include everything that we can imagine, fantasize, dream, consciously or not, and therefore is not actually present— and yet by calling it *virtual*, Tisseron highlights its ability to be actualized in the moment, as it might well be made easier by digital material properties. We understand thus that *virtual* defined in this manner has nothing to do with *virtual* in a technological sense. This must be deplored because the technological meaning of the term, contrary to its philosophical meaning, has indeed come to impose itself for twenty years now on common usage, in a simplified, aberrant, and mistaken form—which is the source of many misunderstandings, which one can find summarized in the French Wikipedia entry under "Virtual."[14] The problem is that Serge Tisseron sometimes uses the term *virtual* in the psychic sense (which is his own), sometimes in the simplified technological sense (which confuses virtual and digital), and rarely or never in the precise technological sense (computer simulation), to the point where we quite often no longer know what we're talking about. It is probably taking into account that ambiguity that led Tisseron in a recent work to introduce the improbable distinction between psychic virtuality and computing virtuality.[15] Despite attempts at clarifying which that distinction introduces, the problem remains that if one recognizes the psychoanalytic notion of the *psychically virtual*, it is neither rigorous nor admissible to introduce the obscure notion of *digitally virtual*. From the point of view of technological reality, there is only the digital (in the sense of the 0 and 1 binary) or the virtual (in the sense of the simulation); there is no *digitally virtual*. Such a notion piles confusion onto confusion and manifests as an obscure pleonasm that overvalues the dimension of virtuality in the phenomenon that is digital world.

From the Neometaphysics to the Vulgate of the Real and the Virtual

The invention of graphical interfaces is undoubtedly the most important event in the history of microinformatics. A graphical user interface (GUI) is a human-machine interaction interface that displays pictorial elements on a screen, which can be manipulated using a pointing device such as a mouse or a touch-screen system. To make computers easier to use, Xerox Park researchers created graphical interfaces in the 1970s based on the famous desktop that Tim Mott imagined. Never wanting to lose sight of *real* users, they created a theoretical model of a user, Sally the secretary: seated at a desk, Sally uses paper and types.[16]

Thought of as a *workstation*, the microcomputer then became the new mold into which our practices have flowed. In just a few decades, we all became Sally, seated at screens, typing on keyboards, and still printing paper. But thanks to graphic interfaces, mainly we were able to become computer users without having to become computer scientists. This enabled us to go beyond "the Apollonian dimension of informatics,"[17] that of the brutal "man-machine relationship," which submits us to the order and complexity of the automaton, to access this "Dionysian, playful, friendly, free image," which leads instead to a "subject relationship" where we reencounter the independence of a creative partner of machines.[18] Some big names from the computer industry, such as Apple, have made this a well-known trademark.

If graphical interfaces are more user friendly and more Dionysian, it is precisely because they are visual and form images. This is the image of the desktop (the desk at which Sally works) and windows (the papers that Sally positions and piles up on her desk), but also the folder icon (the folder in which files her papers), and the trash icon (the paper bin), or the wooden library in the iBooks app on an iPhone or iPad (the personal library that we take along with us when, via our smartphones and tablets, we depart from the *workstation* model). From an aesthetic point of view, we could surely certainly criticize the elegance of some of these images (the wooden shelves in the iBooks application, despite their obvious characteristic with respect to the possibilities, do not appeal to designers), but we ought never forget the graphic talent they can also have (for example, today's cult icons designed by Susan Kare in 1983 for the Macintosh).[19]

The images that graphic interfaces give rise to are revolutionary. From a black screen with which we could interact only by entering lines of code made for experts and other "computer nerds,"[20] they lead us to a visual environment ("pictures rather than text commands," PARC researchers used to say) that anyone can handle thanks to windows, icons, menus, and a pointing system.[21] In other words, GUIs transform John von Neumann's invention, this informational computational monster that is a computer, into an anthropomorphic world of images to manipulate. They make us go, as Sherry Turkle points out, from a "culture of calculation toward a culture of simulation,"[22] that is, from a culture of programming to a culture of the *virtual*:

> The lessons of computing today have little to do with calculation and rules; instead they concern simulation, navigation, and interaction. ... Of course, there is still *calculation* going on within the computer, but it is no longer the important or interesting level to think about or interact with. Fifteen years ago, most computer users were limited to typing commands. Today, they use off-the-shelf products to manipulate simulated desktops, draw with simulated paints and brushes, and fly in simulated airplane cockpits.[23]

This is our daily life in the age of microcomputers: we classify virtual folders, turn virtual pages, draw with virtual brushes, shelve our digital books in virtual wooden bookcases, and so on. In each case, *virtual* means *computationally simulated*. It is therefore just as accurate to say that we file computationally simulated folders, turn computationally simulated pages, draw with computationally simulated brushes, shelve our digital books in computationally simulated wooden bookcases, and so on. As a result, the new images born from GUIs in the 1980s—these computer-generated images that simulate all kinds of (existing or nonexistent) realities—are indeed virtual images, that is, in the computer sense of the term, simulation images.

This is what would lead early digital technology thinkers to embrace the concept of the virtual and, in attempting to merge the computer sense of the term (simulation) with its old philosophical meaning (potential), to give it a second philosophical life in the form of a metaphysical image full of confusion and misunderstanding. The first to take this path, though with a theoretical subtlety that does him credit and whom many did not think to take into account as they built grotesquely on his ideas, is Philippe Quéau, an engineer from the École polytechnique. Quéau's work is informed by many philosophical references. His essays, published by Jean-Claude Beaune in the Milieux series of the Éditions Champ Vallon, skillfully combine philosophical speculation and scientific accuracy. In 1986, two years after the Macintosh's release, while Quéau was research director at the Institut national de l'audiovisuel in Paris, *Éloge de la simulation* appears; it deals with "image synthesizing"[24] and already contains all of the author's major themes. But it is especially with *Le Virtuel: Virtues et vertiges* in 1993[25] that his ideas take on their full magnitude. In this pioneering book, Quéau tries to analyze the philosophical significance of what he introduces as "one of the most recent and most more promising developments in infographics,"[26] namely, "computer-generated images" or "virtual images" technology. When he writes these words, virtual images are already more than they were at the time of the first graphic interfaces from the early 1980s. By integrating "total stereoscopic vision" (vision in relief), obtained by means of "a visual headset equipped with two miniature LCD screens placed in front of each of two eyes,"[27] they have become real, immersive, visual environments. Understand that artificial spaces are synthesized computationally where humans can instantiate or incarnate themselves: "We circled around images, now we'll circulate inside the images. ... Virtual images are not ever only pictures, just pictures; they have underneaths, behinds, belows, and beyonds. They form entire worlds."[28] Quéau rightly calls these entirely computationally simulated worlds of interactive images "virtual worlds" and offers the following definition: "A virtual world is an interactive, graphical database that is searchable and viewable in real time as three-dimensional synthesized images so as to give the feeling of immersion in the image. At their most complex, virtual environments are real 'synthesized spaces' where one can experience the sensation of moving 'physically.'"[29]

In other words, in accordance with the computer meaning of the term, the virtual according to Quéau is none other than the set of "three-dimensional synthesized images that [are] computer-simulated" insofar as they form navigable worlds, that is, spaces that we can move in and live in, so to speak: "Technologies of the virtual summon the spectator's or actor's body to enter the simulated space."[30] Beyond flight simulators, today we can encounter examples of this in realistic universes for scientific use such as Google Earth or fictional and game worlds such as *Second Life*, a virtual world on the Web where users embody virtual characters, create relationships with others, have a social life, and create for themselves a world in which they want to live.[31] According to Philippe Quéau, this "radical revolution of the status of the image in our civilization"[32] can be compared to that of printing or photography. It is an extraordinary opportunity not only for scientific exploration but also for artistic creation. This is what leads him very early on to supporting the potentialities of digital art, which he defines, in the image of virtual worlds, as an "intermediate" art.[33] Using this word, which needs to be taken into full consideration, brings back a whole ancient metaphysics, which comes from Plato and allows Quéau to situate virtual worlds at a very particular ontological level—which is not without its own share of problems. In Plato, in fact, the "intermediate" realities are, on the scale of human beings, realities located halfway between the sensible things of the material world and intelligible forms of the world of ideas: this is about numbers and mathematical idealities. What a fortunate theoretical coincidence! All Philippe Quéau has left to do is assimilate virtual phenomena, which are nothing other than computed information, that is, made of Number, to Platonic intermediate realities:

> Three-dimensional *virtual* images are not analog representations of an already existing reality; they are computer simulations of new realities. These simulations are purely symbolic, and can not be considered phenomena representing any true reality, but rather, like artificial windows giving us access to an intermediate world, in Plato's sense, to a universe of rational beings, in Aristotle's sense.[34]

By *symbolic*, Quéau means that these simulations are a matter of logo-mathematical symbols, that is, languages. Plato's thesis is clear: virtual images are not a *real reality*; they belong to an *intermediate world*. They are somehow floating realities, located between the material and the immaterial world. Neometaphysical reasoning produces its full effect here. But, dare we ask, what would cause us to be subjected to this (even authoritative) ontological partitioning of the world, which emerges from Book VI of Plato's *Republic* and the famous cave allegory in Book VII?

Intermediate worlds do not exist; they are just a metaphysician's fantasy. Nothing serious can be built on them to attempt to understand digital phenomena philosophically even when reduced to a virtual phenomenon. To philosophically understand the digital phenomenon is not to search for similar metaphysical concepts in the history of philosophy, but to try to formulate the philosophy that it contains as a phenomenon of the world, just as the physicist's devices, according to Bachelard, bear philosophical theories.

This neometaphysics of the image leads its author, despite all his efforts to avoid it, to separate virtual phenomena from real phenomena:

> Symbolic representations have more tangible cognitive significance than the realities they are meant to represent. They have a life of their own, which increases on its own by hybridization, confrontation, and recurrent return.[35]

If virtual images have "a life of their own," they form an ontologically separated world and possess, like the intelligible forms of Plato, a distinct species of reality: "The virtual becomes a world of its own, next to the real world."[36]

Philippe Quéau could not be clearer. In this brief statement, he unambiguously formulates the axiom of the neometaphysics of the virtual, which comes to light in the 1990s and which will do much damage. It is about the Platonic belief, so vivid in Western culture, in the existence of a world that is separate from the visible world and would be embodied in the computer age in virtual worlds. Quéau's point is often more subtle:

> It will become more and more difficult to distinguish what is really real and what is virtual because the virtual seems intended to become a hybrid with the real, to constitute a kind of complex real-virtual reality, a new composite reality. The virtual is not outside the real but linked to the real, to make possible what is potential in the real, and to make it happen. The virtual enables the birth of reality.[37]

But the damage is done. When we wake up the Platonic imaginary that sleeps in Western man, this *Hinterworld* imaginary about which Nietzsche has shown that Christianity was only a new iteration, it is very difficult to go back. The poison of belief is so powerful that it is able, as Bachelard taught us, to penetrate the intimacy of the scientific spirit.

Now, under the effect of a new metaphysics of the image, the digital phenomenon will be reduced to a virtual phenomenon and the virtual phenomenon will be considered a "neo-reality"[38] located outside reality. So much for nuances! Against any scientific rigor and despite Pierre Lévy's warnings,[39] the *virtual* from now on will have to mean the opposite of the *real*. And everything that stems from virtual worlds will be considered illusion and fantasy, mirage and deception. Such is the vulgate of the real and the virtual, this imaginary metaphysics that settles in the minds starting in the 1990s, agitating media (and sometimes researchers) at the expense of objective thinking.

The End of the Reverie: "Seeing Things on the Screen at (Inter)Face Value"

While it is true that computers produce virtual worlds, all do not have the same degree of virtuality. The desktop environment of an operating system, the paginated environment of a word processor, the haptic environment of a mobile app, the immersive environment of a virtual world, or the persistent environment of an online role-playing game can all be considered virtual environments, but, strictly speaking, only the last two are, that is, simulated worlds in which users can themselves instantiate themselves as virtual beings (for example, as a character). For the others, one can appropriately speak only of virtual interactive environments.

Since all digital devices are equipped with graphical interfaces, we have dealt for at least the past thirty years with a minimum amount of virtuality (that is, simulation), and some of us more than others. Virtuality is an integral part of the contemporary world's ontophany influenced by digital devices. Does that mean that we have been living for thirty years in a world of unreality, as the vulgate of the real and the virtual would suggest? Those who believe are obviously victims of a fantasy that acts on them as an epistemological obstacle. Trapped by the Platonic metaphysics of the image, they are prisoners of "first impressions, sympathetic attractions, and careless reveries,"[40] from which the true scientific spirit, concerned never to give up the "psychoanalysis of objective knowledge," must turn away. More than any other, the virtual is that thing one must be wary of and for which Bachelard's words, once again, sound so right:

> Sometimes we stand in wonder at a chosen object; we build up hypotheses and reveries; in this way we form convictions which have all of the appearance of true knowledge. But the initial source is impure: the first impression is not a fundamental truth. In point of fact, scientific objectivity is only possible if one has broken first with the immediate object, if one has refused to yield to the seduction of the initial choice, if one has checked and contradicted the thoughts which arise from one's first observation.[41]

To understand the digital phenomenon, you have to break with the first observation. What holds for Bachelard's idolatry of fire holds here for the idolatry of the virtual. The virtual as a (pseudo-)scientific object is built in this "zone that is only partially objective in which personal intuitions and scientific experiments are intermingled,"[42] guiding "the naive soul" that lies dormant in each of us to triumph too often over scientific rigor. For "even the scientist himself, when not practicing his specialty, returns to the primitive scale of values."[43] Caught within archaic beliefs, aftershocks of Plato, researchers themselves sometimes give in to the vulgate of the real and the virtual. And nobody ought to be surprised if, despite the persistence of some enlightened authors who denounce it,[44] we continue today, in research in the humanities and social sciences, to use the rhetoric of the real and the virtual as if it were objective—because, says Bachelard again, "reverie takes on the same primitive themes time and again, and always operates as it would in primitive minds, and this in spite of the successes of systematic thought."[45] The dream of the virtual, anchored in a neo-Platonist metaphysics of the image, is at times the most powerful and the most aberrant of the past twenty years. It deceives us by flattering our metaphysical inclinations and distracts us from objectivity. Destroying it in the hope of learning about digital phenomena is made all the easier today by our twenty-year habituation to interfaces.

Indeed, if contact with simulated digital realities could seem surreal at the time of Philip Quéau, that is no longer the case at all today. Just as at the beginning of the twentieth century we became accustomed to telephonic ontophany ("speaking to each other without seeing each other"), we are now accustomed to digital ontophany: we have learned to live with computer-simulated realities and to consider them things among things. Whether graphic, such as icons, buttons, or avatars; dynamic, such as cut-and-paste actions, undo, upload or download; or narrative, such as the characters of video games and immersive landscapes, virtual realities too have become banal and ordinary things.

In our "simulation culture," Sherry Turkle writes, we are "increasingly comfortable with substituting representations of reality for the real,"[46] that is, with taking simulated realities for realities:

> We use a Macintosh-style "'desktop" as well as one on four legs. We join virtual communities that exist only among people connected on computer networks as well as communities in which we are physically present. We come to question simple distinctions between real and artificial. In what sense should one consider a computer screen desktop less real than any other? The screen desktop I am currently using has a folder on it labeled "Professional Life." It contains my business correspondence, date book, and telephone directory. Another folder, labeled "Courses," contains syllabuses, reading assignments, class lists, and lecture notes.

A third, "Current Work" contains my research notes and this book's drafts. I feel no sense of unreality in my relationship with any of these objects. The culture of simulation encourages me to take what I see on the screen at (inter)face value. In the culture of simulation, if it works for you, it has all the reality it needs.[47]

"To take what I see on the screen at (inter) face value": this is an expression of great philosophical depth. To take something at face value, first, means to consider that a thing is exactly as it appears, the way in which a something is really what it is. The expression comes from the idea that the value of a coin is exactly equal to the amount displayed on its face, head or tail. Taking things at face value means taking for real what is displayed.

When Sherry Turkle writes, "We have learned to take things at (inter)face value," she is putting forward a proposition of great philosophical complexity: we have learned to consider as objects the things that appear on our screens. In a dazzling formula, we are back to full phenomeno-technology. Seeing things at (inter)face value is precisely seeing them as the interface gives them to us to see. Digital interfaces are indeed a new ontophanic matrix, a new form into which our perception flows just as machines were before, in the first and second, mechanized and industrial, technological systems. In becoming part of our experience-of-the-world, they create a new phenomenological perspective by which the virtual beings of the digital technological system occur as phenomena of the world. Interfaces are the new devices that are forging a new coming-into-appearance, as it were. By living with them, we acquire digital phenomenality, and we educate ourselves in the new ontophany.

That is why, thirty years after they were born, virtual images have definitely lost their metaphysical aura. Having become commonplace, they are part of our most ordinary practices: "The use of the term *cyberspace* to describe virtual worlds grew out of the world of science fiction, but for many of us, cyberspace is now part of the routines of everyday life."[48]

Sending messages, shopping online, sharing on Twitter—all of this does not resonate for us any more as being practices in cyberspace, but rather as practices in the same space as the space of the world. The term *cyberspace* comes from the world of science fiction[49] and is already a dated and phenomenologically expired concept, anchored in a reverie of the virtual and the metaphysical imaginary that it bears.[50] It made sense when we perceived digital phenomena as imaginary and unreal worlds, when we could still create email addresses like cyberprof@voila.fr as a young colleague did in the late 1990s, with whom I skeptically learned about the use of electronic mail. Having and using an inbox at the time seemed like a way to enter another dimension, another reality. It would not occur to anyone today, except if he or she were in a furtive or humorous mood, to create an email address like that one. On the contrary, having espoused the phenomeno-technological dynamics of digital acculturation, our email addresses have embraced the new banality of the world and tend to look like vial.stephane@gmail.com. This simple example, from among the semantic practices of our emails, illustrates how much we have learned to see things from an interface perspective, which is to say to perceive things in a new way, to acquire a new way of being-in-the-world.

We have emerged from the reverie of the virtual. Today, we no longer feel that we are being projected into virtual worlds, but rather that we live among digital interfaces. We also use the term *digital* more readily than the term *virtual* because we intuitively recognize a little more of the technological objectivity and the fully real materiality of the computer phenomenon.

In this sense, the hypothesis of virtuality was a first step in understanding the new ontophany brought on by the digital technological system. It made it possible to show that digital phenomena are computer-simulated facts and that the phenomeno-technology of digital interfaces is the one that has us living in a certain ambient virtuality. But because it is never completely released from the grip of the metaphysical imaginary of the unreal, this path of research is now coming to an end and is insufficient on its own to account for the complexity of digital ontophany. The virtuality in which we are living is but one aspect among others of a generalized phenomeno-technology brought on by digital devices. It is time to analyze digital ontophany in all of its phenomenological complexity.

5 Digital Ontophany

A machine does not live or die: it works and breaks down.
—Jean-Claude Beaune, *La Technologie*

The ontophany carried by the digital technological system is unheard of. Whereas the mechanical revolution struck society with violence, the digital revolution strikes with phenomenological violence. This revolution is so huge that it struck thinkers with the same intensity as it hit the people. Even if very early researchers seized the question, they were hypnotized by the new phenomenology of images and perceptual vertigo of simulation—and some still are. That's what led them at first to focus all their research on the hypothesis of the virtual, with the reverie of the unreal that it induces. In order to contain the phenomenological violence of this new ontophany, it was doubtless necessary to push it back out of the real world temporarily. Anyway, we ended up literally confusing the phenomenon of the digital with the phenomenon of the virtual, as evidenced by the definition you still find today at Wikipedia.fr: the term *virtual* is used to "designate what happens in a computer or on the internet, that is to say in a *digital world* as opposed to a *physical world.*"[1]

Now, if it is true that the virtual is an undeniable characteristic of digital devices—inasmuch as they produce computer-simulated beings—it is only one characteristic among others, for digital phenomena are not just about images. Only metaphysicians who are victims of their imaginary have wanted to tackle the phenomenon by giving such depth to the visible. If we turn our attention to designers who are responsible for the design of digital devices, we can see that there lie many other features of the phenomenon beyond the virtual that are just as unheard of and must be taken into account philosophically. This is what a philosophy of technology supported by a technical and design culture is interested in being able to show.

In truth, few are the phenomena given to our perception that require of us such an effort of reason. Usually we just extend our senses toward objects, and they give themselves up in their natural aura without concealment. With digital phenomena, however, the senses have never before been so misleading. Hypnotized by the screens' world of images, our eyes hide from us the true nature of digital beings. They have us believing in things that would be halfway between being and nothingness, things that would neither be wholly real nor wholly imagined. But it is not a question of situating the phenomenon of the digital between being and nothingness, but rather exactly where it is: between being and the screen. Because if it is more complex than a discrete series of 0s and 1s executed electronically on a silicon chip, it is also more subtle than a series of virtual images that would scroll before our eyes as if shadows of Plato's cave.

The dimension of virtuality, in fact, is only one dimension among others that we experience with digital devices. In order to seize the phenomenological complexity of the digital, moreover, in order to shed light on the meaning of what we experience (and who we are?) when facing these interfaces, we need new concepts. We propose here to analyze digital ontophany through eleven categories. They must be considered phenomenological concepts, less with the purpose of describing the phenomenon of the digital as it stands objectively from a technological and scientific point of view (although this point of view is not absent), instead to reveal what it is made of subjectively from an ontophanic point of view, that is, from the point of view of its singular, phenomenal manifestation, as a phenomenon of the "lifeworld" (Edmund Husserl) by the subject.

"Noumenality": The Digital Phenomenon Is a Noumenon

Some phenomena in the world are not properly phenomena in the sense that they do not appear; they do not make themselves known to us through our senses. Located in a perceptual beyond, they are invisible and might nearly make us believe that they do not exist. Take, for example, as we saw with Bachelard, quantum phenomena or, rather, should we say, quantum noumena because that's what defines a *noumenon* since Kant: its capacity to locate itself outside the field of possible experience. And that's what happens with the quantum world, this "hidden world"[2] for physicists, whose essence is mathematical and which for this very reason can never be present before our eyes. This pure noumenality is not specific to quantum processes, of course. According to Gilles-Gaston Granger, it applies to all phenomena in science. It can be defined, in a philosophical sense quite distinct from its computer sense, as *virtual*.

Indeed, in a book of impeccable rigor[3] that all the metaphysicians of the unreal should meditate on, Granger defines three regimes of reality: the probable, the possible, and the virtual. In keeping with the Aristotelian tradition, they are all opposed "not to the real, but to the actual,"[4] defined as "that aspect of reality that one grasps as imposing itself on our sensory experience, or on how we think of the world, as a singular existence here and now."[5] The actual is the here and now of presence, as a sensible presence (a little like Walter Benjamin's aura). From this perspective, "the virtual would be the name given to the non-actual considered essentially and properly in and of itself, seen from its negative state, without considering its relationship to the actual."[6] The possible then becomes "the non-actual in relation to the actual" while the probable is "an actuality fully and concretely viewed in relation to the actual, so to speak as a pre-actuality."[7]

By suggesting these new categories, which he uses as metaconcepts to rephrase "some classic epistemological problems having to do with knowledge,"[8] Granger is well aware of the "natural collusion of the imaginary and the virtual."[9] This collusion, which is at the root of the reverie of the virtual about which we wrote earlier, can be seen in the literary use of the concept, in Borges's short story entitled "The Library of Babel," for example, or again, in some way, in the very particular use Serge Tisseron makes of the notion of the virtual. But for Granger, this is quite rightly "pseudo-virtual," which should not distract us from the "cognitive virtual," that is, fully freed from "this emotional halo which bears its aesthetic value."[10]

Then, because the real in science is always a reality that is explained, Granger suggests defining scientific explanations as "diagrams or abstract models of reality established through thought-detours, through [what we call] virtualities."[11] It's about highlighting the "role of the virtual, point zero of the possible, in the construction of the scientific object":[12] "We would like to show that all scientific knowledge ultimately and basically rests on what we call the "virtual."[13]

From there, Granger envisions "mathematics as the kingdom of the virtual," in the sense that mathematical beings are "essentially abstract," "and not realized as such in sensory experience,"[14] and not related to currency: "mathematics is indeed the science of virtual forms (whether possible or not) of thinkable objects in general."[15] Does it follow that mathematical beings have no connection to reality? Absolutely not. For Gilles-Gaston Granger, and we cannot make too much of this, one has to take the measure of the fact that "reality exceeds actuality and, as we understand it, it necessarily includes some virtuality."[16] Modern scientific objects are made of this virtuality, not just mathematics: "The empirical sciences are also virtual sciences,"[17] except that, thanks to virtualities, they manage to represent the world as it is currently experienced by the senses, following "a relationship principle of the virtual to the actual" thanks to which the empirical sciences also incorporate the probable:

> Any science of empiricism finalizes, by moving into the actual, knowledge that has unfolded in the virtual, thereby enabling an encounter with experience. But at the level of representation, before any effective experimentation, this transition to the actual is prepared thanks to the probable.[18]

From these developments, which deserve a great deal more in-depth exploration, let's remember this paradox: science aims at reality, but it unfolds in the realm of nonactual. This, Granger finally wrote, cannot fail to make us question "the global meaning of our experience of the world."[19] Indeed, he asks, "Is reality reducible to some actualities?"[20] Certainly not. The time has come to stretch the concept of reality: "So we must say that the concept of reality is constructed with a facet of the actual and a facet made of the virtual and the probable."[21]

The best example of this is the quantum *realities* of nuclear physics. Bachelard defined them as noumena. Granger sees in them "an enigmatic and wonderful example of a virtual reality."[22] These are indeed realities perfectly attested to by science, but perfectly inaccessible to perception. It is in this properly philosophical sense that we can follow on Granger in talking about virtual phenomena, even if we would rather call them noumena to avoid any confusion with the computer meaning of virtual. Seen this way, a noumenon is a phenomenon without phenomenality: which does not become phenomenal, which does not manifest, which we does not make an appearance, which does not make it to the experienced world, in short, that we do not perceive—a kind of unmanifested manifestation. The quantum phenomena of physics are such noumena.

But they are not alone: digital processes also fall under the rubric of noumena. Just like quantum processes, digital processes require technological equipment to appear to us. This equipment is the interface. Whether graphic (visual modality), in command lines (textual modality), or haptic (gestural modality), interfaces are the apparatuses of digital appearance; they make possible the phenomenalization of digital noumena and make them visible and perceptible phenomena, in the form, for example, of virtual environments or virtual worlds—in the computer science sense of the term—which so struck the first thinkers of this technology. Therefore, a digital phenomenon is not primarily a phenomenon; it is a noumenon. Just as the quantum noumenon becomes a phenomenon only thanks to the apparatus known as a particle accelerator, the digital noumenon becomes a phenomenon only through the apparatus of computerized interfaces, which are phenomeno-technological intermediaries between the noumenal (mathematical) scale of computational information and the phenomenal (sensory) scale of the user interface.

So just as nuclear physics brought about the birth of quantum beings, the digital technological system gave birth to digital beings. This is not just an extension of the field of ontology but of the field of the materialogy. A new material is born, with unknown properties, and is, unlike quantum matter, put into everyone's hands in record time. Every technological revolution is a materials revolution; the digital revolution is a revolution of computational matter[23] and, today, is the basis of the "algorithmic medium."[24] Just like the cast iron and iron of yesteryear, computational matter is now available at low cost and in very large quantities accessible to everyone. That's why it shapes our world. As Paul Mathias points out, digital flows are not just in the world, they *are* the world:

> Information is the real—not in the real or in front of the real or near the real or after the real. *It is the real* means that the organization of life as a whole ... is as if infused with information flows, which are not its simple tools but its architecture and actual dynamics. ... Networks are not in reality, it is reality that, on the contrary, seeps out of the confluence of networks.[25]

Networks create the real or make the world. Such is the digital type of phenomenological casting process in which we live. Computational matter is its foundation. It circulates at all scales of life and carries, in addition to multiple uses, a new phenomenology of the world. Only a new material can engender new perceptive modalities. But because its essence is mathematical, that is to say imperceptible, computational matter is, in the first place, noumenal.

Digital beings are noumena—that is, the primary characteristic of digital ontophany.

Ideality: The Digital Phenomenon Is Programmable

In introducing simulation culture, graphic interfaces have opened up a new world. Thanks to the intuitiveness of icons and buttons, they made it possible for all those who do not know how to write code to make use of what computational matter has to offer. Yet we must never forget that behind every icon and every button, there exist lines of code. Computational matter is, first of all, a set of idealities or rational beings, and all about programming languages. Certainly, for an exceptional mind such as John von Neumann's, "the most perfect language and the most universal [is] machine language,"[26] a language that is exclusively made up of a discrete sequence of 0s and 1s. But since only a few exceptional individuals are able to read and understand it, developing other languages was necessary, which, although still very abstract, could be read more easily by most people. Thus were born programming languages, which can be defined as formal languages, consisting of symbols and reducing a problem to an algorithm. The first to have written a programming language is Alan Turing (to whom we owe the modern concept of an algorithm), in order to facilitate the use of the Manchester Mark I, the first computer, built in 1948. Containing only fifty instructions, it strongly inspired Eckert and Mauchly's UNIVAC 1 of 1951. But the first true programming language, developed between 1953 and 1956 for the IBM 701, was Fortran[27]—the language that the first nonexpert users, architects, designers, engineers, physicists, learned to manipulate when GUIs did not yet exist.[28]

Nevertheless, just as human beings do not know how to read machine language, machines do not know how to directly interpret programming languages written by humans. So we invented a particular type of language used only to translate a programming language into a machine language (or another programming language). This type of language is called a compiler. Thus, at each moment in a computer, there is a series of chain compilations allowing high-level languages (which have the highest degree of abstraction and can be read by humans) to be translated into low-level languages (whose syntax is closer to the machine's binary code), up to machine language itself. These nonstop operations take place at all levels, from command line interfaces to GUIs.

For example, when I type these lines in my word processor (software written in high-level language), a series of instant compilations takes place allowing my action, executed in machine language in the electromathematical limbo of the microprocessor, to produce an immediate text display on my screen. Philippe Quéau very early underlined this characteristic when writing about virtual images: "Unlike photographic or video images that stem from the interaction of real light with photosensitive surfaces, these images are not originally images; they are in the first place languages:[29]

> Computer-generated images are computed from mathematical models and various data. This is called *image synthesizing* because all of the information needed to create an image or even a series of animated images is available in symbolic form in the computer's memory, and so there is no need to appeal to the *real* world to create them. ... The readable can now generate the visible.[30]

A virtual image, then, is nothing more than a "computed image,"[31] and "the deep nature of the virtual is of the order of writing."[32] Twenty years later, Paul Mathias similarly writes, "The Internet is a world. Not of things, machines, and instruments, but of meanings. ... The Internet is an infinite process transitive and transitory writing."[33]

As a result "virtual worlds [are] totally synthetic; they can be programmed at will."[34] But what is true for virtual environments, as we now understand, is only a general characteristic of digital beings, whether or not they include virtual images. Everything a computer does, whether a large system computer, a microcomputer, a Web server, or a touch pad, is generated by lines of code, that is, by programming languages. We live in the time of software or, rather, as Lev Manovich says, alluding subtly to S. Giedion, a time when "software takes command."[35] In this sense, even if we understand its descriptive value, we might doubt the relevance concept of the term *artificial intelligence*, which tends to suggest that a computer would have a natural autonomous intelligence, without it being in one way or another a consequence of a series of instructions programmed and wanted by human beings.[36] Sylvie Leleu-Merviel points out, "Even though computers indeed produce propositions actually understandable by humans, they are by no means *intelligent* in the sense that they have no *understanding* of anything they do."[37] Computer scientists are well aware of this: that nothing that happens within or using a computer is caused by anything other than a program written by one or several human beings. Digital beings are by definition programmable phenomena.

That's why programmers—whom we call developers today, or hackers in the best sense—are the heroes of our time. Just like the Renaissance engineer, a hacker embodies today's figure of eclectic genius. From Steve Wozniak to Bill Gates, from Richard Stallman to Tim Berners-Lee, from Linus Torvalds to Steve Jobs, it is computer scientists who make the world in which we live, as well as those who, in the daily life of the creative industries, create the interfaces we use every day (software, websites, applications). To work with computational matter is a demanding operation of the mind, as much of the "mathematical mind" as of the "intuitive mind" (Pascal). We should therefore take Pierre Lévy very seriously when he invites us to consider "programming ... as one of the fine arts":[38]

> In the profession of computer science, there is a whole lot of creativity and inventive cooperation, which is generally poorly known by the general public. ... At the time of the Renaissance, it was normal for artists to be surveyors or engineers, humanists. Alas, since that time, disciplinary specialization has devastated the intellectual landscape. While the job of computer scientists is to arrange architectures of signs, to compose a communications and thinking environment for groups of humans, we strangely refuse to consider their activity as informed by artistic and cultural competence.[39]

Those who work with computers know this. When Steve Jobs was asked to explain the Macintosh's success, he replied that, above all, "the people who worked on it were musicians, poets, artists, zoologists, and historians who happened also to be the best computer scientists of the world."[40] And for sure, the love of code in computer culture is as strong as the love of art in classical culture: "Code is poetry," WordPress's official slogan. Just as when Steve Wozniak—dubbed Wizard of Woz—managed to get eight chips on an integrated circuit to do the work of thirty-five chips, he was committing not just an act of technical genius but also an act of love:

And I took this book home that described the PDP 8 computer[41] and it just ... oh, it was just like a bible to me. I mean, all these things that for some reason I'd fallen in love with, like you might fall in love with a card game called Magic, or you might fall in love with doing crossword puzzles or something else, or playing a musical instrument, I fell in love with these little descriptions of computers on their insides, and it was a little mathematics, I could work out some problems on paper and solve it and see how it's done, and I could come up with my own solutions and feel good inside.[42]

We know the future of one of his "own solutions": it bears the name Apple II, the first massively marketed consumer microcomputer:

And then I got in to a way of why have memory for your TV screen and memory for your computer, make them one, and that shrunk the chips down, and I shrunk the chips here, and why not take all these timing circuits and I looked through manuals and found a chip that did it in one chip instead of five, and reduced that, and one thing after another after another happened. I wound up with so few chips, when I was done I said hey, a computer that you could program to generate colored patterns on a screen, or data or words or play games or anything, it was just the computer I wanted, you know, for myself pretty much, but it had turned out so good. He said I think we have a computer we could sell a thousand a month of.[43]

In addition, just like artists or scientists, computer scientists are never more themselves than when they endow their code with the power to carry a vision of world. When he chose to place the World Wide Web in the public domain in 1993, Tim Berners-Lee did not simply commit an act of sharing technology, but a political act as well: he chose to offer to the world, without applying for a patent, the most powerful hypertext system that has ever existed. We know what happened to these few lines of code assembled in a certain order!

So we can say that all that is digital is abstract and semantic. Computational matter is of a logical nature, in the sense that it consists of mathematically and formally organized rational beings, of signs, information.

The digital phenomenon is therefore idealism. Such is the second characteristic of digital ontophany.

Interactivity: The Digital Phenomenon Is an Interaction

In 1979, while Steve Jobs has the opportunity to visit Xerox Park where he learns about the Xerox Alto's graphical interface, Bill Moggridge is chosen to design the first laptop, the GRiD Compass, which was marketed in 1982 and then traveled on the space shuttle *Discovery* in 1985. "It was a huge thrill to be a member of a team that was in the process of doing something so innovative," he says in a video interview.[44] In 1981, while Apple is working on the Macintosh, the first functional GRiD Compass prototype is finished. Dedicated to very specific applications, it has no GUI, yet it is a major breakthrough incorporating a number of innovations. The most famous is the flip screen that shuts down the computer when it is closed. In carrying out this project, Bill Moggridge understands that with digital technologies, one must rethink the entire design process to be in the service of the user.

In 1986 with former Xerox Park researcher Bill Verplank, Moggridge forges the designation *interaction design* to replace *user interface design*, which human-computer interaction (HCI) engineers used until then. Hence, industrial designers today tend to consider the term *interface* outmoded and obsolete, which, they say, belongs to a past period of design culture, whereas Web designers used to talking about *UI design* still deem it relevant to characterize the purpose of their work. For Moggridge and Verplank, moving from an interface vocabulary to one of interaction has a purpose: to focus on the experience rather than on the object, and thus to move from an exclusively technical culture to a design culture: "Product designers of digital technologies products no longer consider their work to be to design a beautiful or useful physical object, but rather to design interactions with it."[45]

Interaction emerges as a new notion. And it comes not from philosophers but from designers. It emphasizes an essential characteristic of digital phenomena: since digital devices can be used only by means of interfaces (whether they are volumetric, software, visual, haptic, or gestural), the nature of the operational experience they provide to the user is not of the order of mechanical action but of algorithmic interaction. When using an object that is not computerized, a typewriter, for example, one can say that one acts, in the sense that one produces a mechanical action with one's body (we press a key), which has a direct impact on the machine's material (its gears and levers), itself resulting in a physical action (typing a character on the ink ribbon). But when one uses a computer, one does not act: one interacts.

First, because our body cannot come into direct contact with computational matter (half-mathematical, half-electronic, that is, noumenal, inaccessible to perception through the senses), we must instead use hardware and software substitutes (keyboard, mouse, icons, buttons, menus)—which, precisely, we call *interfaces*—making this matter perceptible, manipulable, and usable for all kinds of purposes (playing, working, buying, selling, chatting, communicating). These phenomeno-technological intermediaries transform the digital *noumenon* into a digital *phenomenon* and thereby establish a junction between computational matter and us, emphasizing at the same time our irremediable separation from it. It is an edifying paradox, which teaches us that living among interactions (or experiencing an interface) is to live between two worlds, as in suspension. Others call this "life on screen."[46]

Then, and this is the second reason, digital material is reactive: a user's action leads to the system's reaction, as if the machine responded to us and engaged with us in a relationship or a dialogue, which we prefer to call an interactive situation. The microcomputing pioneers speak about this, as did Steve Jobs here:

> So you would keyboard these commands in and then you would wait for a while and then the thing would go dadadadadada, and it would tell you something but even with that it was still remarkable—especially for a ten year old, that you could write a programin BASIC let's say or FORTRAN and actually this machine would sort of take your idea and it would sort of execute your idea and give you back some results and if they were the results that you predicted your program really worked it was an incredibly thrilling experience.[47]

In other words, the user acts, and the machine reacts. I click on a button, and the title of my text is displayed in bold; I click on a link, and my Web browser takes me to the requested page; or I press on several keys, and my game console makes me experience driving a car. Therein lies the interactivity because interacting is precisely reacting to a reaction, which causes a new reaction to which we must react again. ... Living amid interactions is thus engaging in a potentially infinite relationship with computational matter, as if it were an interlocutor who always *gives something back*. That's why, to the chagrin of the worried, digital interfaces are very powerful attention grabbers: they endlessly seek our ability to interact with them. But that's what makes them equally attractive and *play*sant, as it were—which we will come back to.

Thus, the digital noumenon is not an inaccessible phenomenon reserved for only the Beings of Reason of mathematicians. Because computational matter is both programmable and programmed, it is basically reactive, that is, accessible to a user. We must understand that this reactive ability is one of its intrinsic properties and that it is only because this property is intrinsic that we are able to speak of interactivity. Otherwise, everything that is an intermediary between two things could be an interface; such an abuse of language is not acceptable, even if interaction designers themselves are sometimes guilty of it.

Rigorously speaking, only that which in its very materiality contains an intrinsic ability to react can claim to be an interface. A movie screen is not an interface and produces no interactivity. Of course, one will object that sitting in front of a filmed projection, I am never passive because I feel an intense activity within me in the form of representations and emotions. But in this case, it is I who reacts, and I alone, with my own psychic matter, in summoning my imagination, my unconscious, my fantasies; it is not the sequential flow of images projected on the screen. The cinematographic images projected on the screen are not capable of reacting technically. They are not interactive, only active; they are mechanically connected, following the irretrievably fixed order of the film without being able to change along the way. We cannot stop the film or alter the script by clicking on a scene. We cannot do that because cinematographic images are not made of programmable material. They are not digital; they are made of photosensitive material, that is, mechanical. That's why they are incapable of generating the least bit of interactivity. A movie theater is a (magnificent) technology of the mechanical age. The same goes for the telephone, which is not an interface but an intermediary; it puts us in touch at a distance, but it does not create more interactivity than when we talk to each other, in our private moments, in bed. It is only a means of transmitting sound, capable of fostering relationships but not interactions.

We should not confuse interactivity with any activity generated within me by an object located outside me. Only what is made of computational matter can generate interactivity, that is, activity correlatively produced by me and by an object located outside me. This is the case of video games, which, unlike cinematographic images, are objects with some of the most powerful, intrinsic, reactive aptitude that has ever existed. A video game is an exemplary technology of the digital age and exploits all its potentialities.

In a video game, not only am I *instantiated* in a virtual world as an interactive element (my avatar), but I also have the actual material possibility of interacting with the virtual world around me, thanks to interface elements such as menus (the options of the game), virtual objects (a weapon, a garment, a automobile), nonvirtual objects (a joystick), and even my bodily gestures (detection of the joystick in the space). In this sense, a video game is a model of interactivity and proves Mathieu Triclot's point when he asks, "What cultural form, other than video games, makes this possible? Who can rewind a movie or turn back the pages of a book and hope that what follows will be modified according to one's expectations?"[48]

This is why interactivity is a sure criterion of demarcation between video games and movies. "Watching is not playing."[49] Only a digital interface can be a substrate of interactivity and, conversely, interactivity is characteristic of digital interfaces.

When a digital phenomenon opens itself to perception, it is basically an interactive phenomenon. That is the third characteristic of digital ontophany.

Virtuality: The Digital Phenomenon Is a Simulation

Although the first thinkers of the digital age wanted to give virtuality a central and essential role, it is, among all the characteristics of the digital phenomenon, the only one that is contingent and accidental. This means that a digital device does not necessarily contain virtuality, at least in the computer sense of the term, whereas it necessarily contains interactivity or programmability.

Command line interfaces, for example, are the utmost digital devices but without any virtual environment. At the most, they can use *virtual memory* (a nonvisual type of simulation). Yet we have long wanted to make of the virtual the essential characteristic of the digital—we need not repeat this. We have shown this: the virtual refers only to the capacity of digital devices with graphical interfaces to produce simulated realities, whether software environments (such as Windows or Mac OS) or real virtual worlds (such as *Second Life*).

We must insist here on the phenomeno-technological aspect of the virtual. By this, we refer to the fact that the inherent virtuality of digital interfaces—which are nearly all graphical interfaces—is the most apparent part of the mold where our perception flows nowadays. Why? Because the virtual is the best visible representative, on a phenomenal scale, of the computational matter that operates invisibly at the *noumenal* scale. Only virtual environments succeed, in addition to providing us with greater ease of use, in embodying the digital noumenon in the field of our perception. Considered from this point of view that virtuality is not only at odds with the reverie of the unreal so dear to contemporary neo-Platonists, but it must be understood, quite to the contrary, as the only way for a digital noumenon to become a phenomenal reality—as if the virtual, to speak falsely as does the vulgate, were the measure of the real, that is, the only way to apprehend or manifest it. This is perhaps why the virtual is nothing other than a simulation: in order to make visible an invisible reality, there's nothing like simulating it, thereby making it a phenomenal instance.

Virtuality is therefore a major phenomeno-technological structure of perception in the era of the digital technological system. That is the fourth characteristic of digital ontophany.

Versatility: The Digital Phenomenon Is Unstable

One day in September 1947, with those working on the Harvard Mark II under Howard Aiken, the presence of a moth in relay 70 of panel F completely interrupts the machine before the incredulous Grace Hopper, mathematician and naval officer, future designer of the first compiler (1951) and of the COBOL language (1959). With the help of tweezers, Grace dislodges the most renowned moth in computer history and glues it in her lab notebook with the caption: "First actual case of bug being found." Even if Thomas Edison had already used the word to designate a flaw, Grace Hopper was the first to apply it to computing. The term quickly established itself to designate any design error that causes the malfunction of a computer program. The concept of debugging that Hopper introduced takes on its full meaning: it is a question of searching the code for a logical parasite that crashes the system.

What happened to the Harvard Mark II will continue to happen to all of the computers that succeed it. Why? Because a computer cannot *live* without bugs. Even though human beings write computer programs, the programs are never a priori fully mastered. For a program to be fully operational, the machine must execute many times to check its behavior in every circumstance and rectify the inevitable gaps. No programmer in the world, whatever that person's talent, is able to write a program that runs from the start without a bug. That's why software, an application, or a website undergoes many tests and debugging before it is launched. And that's why computer scientists always number versions of their programs very carefully, granting 1.0 only to a software program that has been subjected to a great deal of testing. Despite this, as everyone knows, possible loopholes in the program code make it necessary to provide regular updates and the development of *security patches*. It is impossible to write code without generating bugs, if only to correct them all afterward. Bugs are consubstantial with computational matter.

In other words, digital material is necessarily matter that falters, stumbles, or falls. And when a server is the victim of a bug, we say that the server is *down* or for a website's code that the site has *crashed*. And that is not only true for developers. Once it has started its life cycle and is put in the hands of a user, a program always ends up producing a bug. For example, early users of Microsoft's Windows operating system all experienced at least once the famous bug known as the "blue screen of death" that displayed when it encountered a critical error. Likewise, regardless of what operating system is being used, it happens to everyone to have to reboot the machine, relaunch an application, or restart an action. On an iPhone, for example, from time to time an application abruptly closes for no apparent reason, or one cannot receive an incoming call, even though one's finger swipe is just fine.

In computational matter, there is a structural and unpredictable tendency for bugs: this is the versatility of the digital phenomenon. Some programs are known to be more stable than others—GNU/Linux servers, for example. But ultimately, whoever the builder or developer, there will always be an irreducible tendency to instability in a computerized product—not to mention anomalies resulting from malicious acts like viruses or attacks. Floridi recently called this "the unsustainable fragility of the digital."[50] This instability is part of the ontophanic culture we have learned to live with. Accustomed to the functional vagaries of our computers, we now know that they can crash. That is why we perform regular backups.

Living in digital ontophany means living by the side of unstable matter to which we entrust everything, but without ever being able to trust it completely. And we know it. We have learned to live with this instability, sometimes with costly fantasies and delusional fears, such as that of the so-called Year 2000 or Y2K bug.

In truth, we find it difficult to accept that something as powerful is also so fragile. The faster computers get and the more they are connected to high-speed broadband transmission, the more immediately we expect the machine to respond to our expectations. This makes us more and more intolerant of bugs even though they are part of the nature of computational matter. An education in digital versatility is still lacking. We should teach our children to live with bugs, to try to accept their effects, and to bypass any damage—or even any benefit. There can be fortuitous bugs as there is sometimes, in an act of creation, some good fortune (for example, glitch art). Remember the painter Apelles's well-known mishap, reported by Sextus Empiricus:

> It is said that he [Apelles], while he was painting a horse and wanted to imitate in his painting the animal's froth, was so far from his goal that he gave up and threw at his painting the towel with which he wiped the colors from his brush; when it struck, it produced an imitation of the frothing of the horse.[51]

Hence the expression, *throw in the towel*. The computer science bug can be likened to the painter Apelles's towel: a felicitous surprise—unless it is closer to what happens when I try to drive a nail in. Before being able to nail it in straight, I have to take hold of it again, line up the hammer again, and try again. Computational matter is similar: it does not always produce the right effects the first time. Sometimes it needs restarting or relaunching. Because it's unpredictable, it introduces into our experience of the world a phenomenality of the unstable.

The digital phenomenon is versatile. That is the fifth characteristic of digital ontophany.

Reticularity: The Digital Phenomenon Is *Otherphanic*

For twenty years in France, the word *network* has been uttered and written everywhere; it has been analyzed from every possible vantage point;[52] we learn of its unexpected roots;[53] we use it to name scholarly journals;[54] and under the erudite name of *diktyology*, we even try to make it an ontology.[55] Yet it must be said that it is only with the networks that we call *social media*, which appeared during the 2000s, that networks revealed, long after being invented, their full potential.

With the internet it is no longer just about cyberspace and the interconnection of machines (as in the time of the reverie of the virtual), but about social connections and our relationships with others—everyone can take its measure by observing one's own uses on Facebook, Twitter, LinkedIn, or Instagram. The internet, as sociologist Antonio Casilli underlines, "is where we exchange emails, chat with friends, and share music and photos with strangers. And this communication is, precisely, a social fact, helped and shaped by computers,"[56] with the exception that nowadays we do it as much and even a lot more with people we know than people we don't know.

What is of interest here is not so much a sociological analysis of our practices as the philosophical significance of a social bond "shaped by computers." In the era of the digital technological system and interface devices, the social bond does not escape phenomeno-technological constructivism. This means that the modalities of the social bond are historically and technically conditioned a priori. For an individual, the social links that are activatable within a group depend on the devices that allow them to be triggered, by activating them, to phenomenalize them, in a way that bears the ontophanic mark of these devices. According to the technological system in which we live, we do not therefore elaborate the same ontophanic culture of our relations to others, because the devices that need to be mobilized to establish these relationships are not the same ones.

For sure, at the origin of psychic life, in an age when the world did not exist yet, there was no need of an apparatus—other than our psychic apparatus—for a relation to an other to exist. As the French psychoanalyst Serge Lebovici teaches us, not without encountering the common philosophical meaning, "the object is invested in before being perceived."[57] But in adulthood, in the field of social relations (those that, because they are not intimate, are not too inflected by the psychic vestiges of childhood), let us acknowledge that there is always a technological device between others and ourselves, without which it would be impossible to establish contact. Whether we're talking about the table in a restaurant that creates the spatial conditions for a face-to-face encounter, the telephone that creates the sound conditions of speech without face-to-face encounter, or the online social media that create the interactive conditions for a connection that can be both without any words or any face-to-face encounter, the relationship to another is always, in its phenomenality, technologically conditioned. In other words, as a component of my experience of the world, the entirety of the social bond is the fruit of phenomeno-technological casting. Antonio Casilli gets close when he writes:

> The structures of an online society are not limited to faithfully copying the logic and social processes that we can observe once we shut down our computer. The Web promotes new ways of living in society whose impact, because of the omnipresence of networks, ends up going beyond user circles and becoming a sign of our time.[58]

The social connection in the digital age is therefore indeed "shaped by computers" in a way that marks our era. As ontophanic machines, digital machines offer a form into which our relationships flow, just as yesterday's mechanical ones did in other ways.

The reticularity of the digital phenomenon is therefore not just a technological fact of interconnection; it is a phenomeno-technological fact of ontophanization. As the telephone once did, the internet is generating a new ontophany of otherness—what we might call, in a play on words, an *otherphany*. Thanks to the mobile digital interfaces of the 2000s, which make of the network a constant ubiquitous reality, an other is potentially always here, in my pocket, at my fingertips. The problem is not then whether, in our age of digital devices, we suffer from a hyperpresence or whether we have become incapable of solitude.[59] Solitude is above all, as everyone knows, a psychic disposition that can be activated anywhere, including with family or on the subway, in the tightest crowds. The question is more to figure out how long it will take us to finish getting used to the digital ontophany of hyperpresence, just as we became accustomed, in the middle of the twentieth century, to the telephonic ontophany of telepresence. Like the previous ones, the new ontophany of the other will soon become common and natural, without anybody seeing in it any malice or problem.

This is the lesson of the history of technology. It amounts to the fact that the opposition between an online sociability and an offline sociability basically has no reason for being: in all historical periods and in all technological systems, we have always been in-relation-to-others and using phenomeno-technological intermediaries.

There is no such thing as a truer or more false technological modality when it comes to being in relation to others. My friends are no less true or real when I engage them on Facebook than when I dine with them at home. Rather than succumb to the reverie of the virtual, which leads to considering sociability online as unreal, we simply have to accept the notion that our modalities of social interaction are, thanks to digital technologies, augmented with new possibilities, without eliminating or replacing earlier ones. Sociologist Casilli, taking a different path, reaches the same conclusion: "For its users, the sociality of the Internet is no substitute for work relationships, kinship, or friendship. It adds to them. Digital technologies are not therefore a threat to the social bond. They are complementary modalities."[60]

So, depending on the significance or the function we want give to our relationships, we can now invite others into the field of our experience-of-the-world through a varied range of ontophanic modalities: face-to-face dialogue, telephone voice, remote messaging (SMS), public exchange of ideas (Twitter), sharing photographic images (Instagram), social networking (Facebook, LinkedIn), and so forth. In our digital age, we are living a more augmented experience of the world than ever before, in which the possibilities of being in touch with others have never been as rich and varied. If the digital phenomenon is a network phenomenon, then it is above all in the phenomeno-technological sense that it revolutionizes the conditions under which the other appears in our field of experience. The network is an ontophanic matrix of new forms of social connection, the new techno-transcendental structure of sociability.

The digital phenomenon is an *otherphanic* phenomenon. This is the sixth characteristic of digital ontophany.

Instant Reproducibility: The Digital Phenomenon Is Copyable

Although it has unleashed legal and political passions in recent years, the instantaneous reproducibility of digital information does not seem to have caught the eye of thinkers who were doubtless too involved in the reverie of the virtual. This is, however, a feature that is not only unheard of in the history of materials, but also, from a phenomenological point of view, even more prodigious than virtuality. What is it about? Simply the actual technological possibility of instantly generating potentially infinite copies of the same element, image, sound, book, or whatever: for the processor, it is just a series of discrete 0s and 1s.

Each one of us experiences a simple and immediate experience when we send an email. Let us imagine that you have a PDF version of Homer's

Odyssey, and you send it by email to one hundred people. In less than a second, this copy is reproduced one hundred times in full and distributed one hundred times to one hundred recipients, themselves located in one hundred different places on the globe, beyond mountains and oceans. If we had been able to show such power of transmission to Johannes Gutenberg, Peter Schöffer, Nicolas Jenson, and a few other masters of the typographic art of the fifteenth century, who knows whether they would have been exalted or dismayed. Think about how long it took in their era to compose the text of a single page and achieve a few dozen copies many months later.

Today, millions of content downloads occur every day all over the world. The unparalleled ease with which it is possible to reproduce digital content is such that it is impossible for users to give it up. And this is irremediable: we cannot return to the horse-drawn carriage postal service once we have had a taste of email. That is why the redesign of authors' rights is not only desirable but inevitable. What is being played out goes well beyond the financial interests of the creative industries. What is at stake is the possible artistic and cultural experience-of-the-world. Users now have the opportunity to experience art and culture in a way that increases their ways of being to such an extent that they will never look back.

Take the example of the iPod, which some have dared call "the perfect thing."[61] The iPod has changed everything in how we experience music: first, as that pocket-sized object for carrying around our entire personal media library, and then as an embedded iPhone app connected to continual download services, for accessing everywhere on earth the sum total of the music available worldwide. The iPod is therefore a phenomeno-technological device, that is, a form where our experience of culture flows. It transforms our experience-of-the-world by generating a new *ontophany* of music: all music all the time. Music is everywhere, at all times, and the world becomes, so to speak, a musically assisted planet (additional proof, as if it were needed, that music is one of the major arts of our times)—with numerous consequences, in both the private space, where the spatial volume required for storing music has been reduced to a small device one can hold in one's hand, and the urban space, where nomadic listening affects all generations.

On its own, the iPod embodies the ontophanic paradigm of cultural consumption in the era of the digital technological system. Because cultural goods and intellectual work all tend to become networked digital flows, they acquire a new phenomenality, made of lightness, fluidity, and ubiquity. Thanks to that new ontophany of art (and knowledge), not only is our cultural experience of the world enriched and broadened, but also especially simplified and made easier. How many of us, crushed by overloaded schedules, no longer have time to go buy music in some endangered record store, preferring to effortlessly purchase whatever music we like—or podcasts of our favorite radio shows, the best TV reruns, not to mention digital books and other educational content? iPods introduce us to a totally ubiquitous culture. With them, all of the cultural wealth produced becomes available at every instant of our experience of the world. There is no doubt that nothing else as great has happened since the invention of the printing press.

The digital phenomenon produces instantaneous, reproducible beings. Such is the seventh characteristic of digital ontophany.

Reversibility: The Digital Phenomenon Can Be Canceled

Ever since Antiquity, philosophers have not stopped reminding us that we are mortal and that human existence is inherently tragic. The death of living beings is genetically programmed; it is a law of nature. To her son Hamlet, who refuses to accept it, Queen Gertrude tries to remind him: "Thou know'st 'tis common; all that lives must die, Passing through nature to eternity."[62]

All living things must die one day. And that is true not only of living beings. "All things are born and die,"[63] Heraclitus wrote. The entire physical universe is subject to entropy, that is, to increasing disorder. Death is but an illustration on the scale of living things of the fundamental irreversibility of the universe. In a 2009 film, *Whatever Works*, Woody Allen gives it an amusing illustration when he has the main character—that physics genius who failed in his marriage, failed to get a Nobel Prize, and even failed to commit suicide—say that the entropy of the universe, in the end, is that you cannot put the toothpaste back in the tube.

One of the most fascinating ontophanic modalities that digital phenomena introduce into our experience-of-the-world is precisely the possibility of going back—not just like in a movie, where we can rewind the film to watch a scene again that remains mechanically and invariably the same, but rather like in a video game, when we return to a previous step to *start again* and invent new plot lines interactively. Mathieu Triclot correctly points out how much reversing our experience, which is quite new, provides "a specific and gigantic pleasure, that of being able to resume and repeat without hindrance a sequence until it is satisfying."[64] And this pleasure is a digital pleasure: "Video games enable a new pleasure, which is intimately linked to the computer, to the confrontation with a universe generated by computation."[65]

Indeed, before the digital technological system, never had it been possible for a human being to experience anything completely reversible. That is not just the case with video games, which are somehow the maximal form of interactive experience. Everyone is constantly experimenting with it, even if in a minimal way, with everyday interfaces. In the country of computational matter, it is always possible to *cancel*. *Ctrl-Z* or *Command-Z*, Undo, is the best-known two-key stroke of computer science, to which we are already so accustomed that we are sometimes sorry, as if by reflex, not to have it at our disposal in the classical physical world.

No matter what you cancel—a character you typed, an email you sent, the several thousand batch-processed images, the adjustment of a piece of hair you made using photo-editing software—for the processor, it's just about running a series of 0s and 1s among others. But for the user, in the area of phenomenological reception, doing so is big and almost supernatural: it's nothing less than the annulment of the fundamental irreversibility of the physical world to which we belong. That is, without a doubt, one of the particular reasons, characteristic of the technological objectivity of computational matter, that has helped maintain the illusion, under the name *virtual*, that digital worlds were not quite real. Irreversibility is not a natural behavior in accord with the laws of physics. Nothing in the universe is reversible, if not (in appearance) the digital phenomenon. And this digital phenomenon, whatever the phenomenal reception, is an objective physical reality: a sequence of 0s and 1s electronically executed on a silicon chip.

It must be recognized that the digital phenomenon, in offering the field of our experience the possibility of going back on or canceling, introduces a new phenomenality at the same time: the ontophany of reversibility. Contrary to all that humanity has been used to at the perceptual level for centuries, this one is too unreal, as it were, not to provoke in us a phenomenological jolt—which some people experience more or less well. We are not used to things being so contingent. We are rather more accustomed to dealing with an irreducible feature of necessity. Some things, the Stoics taught us, have nothing to do with us and therefore cannot be other than they are. But in the digital world, Stoicism no longer holds. To be sure, things are what they are (they even have their own determinism), but they can at any moment be something other than what they are; they are reversible.

Should we see a danger in any of this? A disturbing opportunity to lose our awareness of the finiteness of existence and of the fundamental irreversibility of being? No. As in all historical periods, humanity is simply confronted with new ontophanic learning. Whatever vertigo digital ontophany brings, it will happen like others before it, and humanity will assimilate the ontophanic culture of reversibility, finding the appropriate distance to locate it within our experience-of-the-world. Certainly, as in a dream of eternity or a desire for immortality, we would sometimes like the whole world to be reversible just as computational matter is.

But only digital phenomena are reversible. This is the eighth characteristic of ontophany digital.

Destructibility: The Digital Phenomenon Can Be Annihilated

In our time of global warming and energy transformations, we understand more than ever how much industrial innovation is not just based on our ability to invent new materials to produce them in large quantities, but maybe rather on our ability, after they are produced, to turn them into waste that can disappear. The case of nuclear materials, whose half-life can reach millions of years, is edifying. From the point of view of a user's *experienced* world, the way computational matter too can naturally disintegrate ought not to leave us indifferent—and isn't this another incredible feature of the digital phenomenon?

For sure, electronic devices and their toxic components comprise a worrisome amount of waste. As John Thackara reminded us in 2005, "The amount of waste matter generated in the manufacture of a single laptop is close to four thousand times its weight on your lap."[66] However, if one focuses on the strictly phenomenal reality of computational matter, one can only be struck by the way it is able to fade away, so to speak, without leaving a trace. A simple hiccup in electrical power is enough for everything not saved in memory to literally and irreversibly disappear from one's field of reality (because the digital phenomenon, reversible as it is, nonetheless remains attached to the fundamental irreversibility of the physical world). Where, then, did all the lines of text that we wrote go, or the images that we were in the process of touching up? They were but a discreet suite of 0s and 1s pending storage. They vanished at the very moment the electric current stopped flowing through the millions of transistors of the microprocessor. They simply vanished.

In principle, there is no material capable, in its physical reality, of disappearing this way, without leaving a trace, by simply fading from the field of reality. We all learned that at school: water brought to the boiling point does not disappear; it turns into steam. Even the text I write with chalk on the blackboard leaves traces: when I erase it, it becomes dust on my fingers. After Anaxagoras of Clazomenae, Lavoisier made this a fundamental principle of physical science: "Nothing is lost, nothing is created, everything is transformed." In a famous section of his *Elements of Chemistry*, he writes:

> We may lay it down as an incontestable axiom, that, in all the operations of art and nature, nothing is created; an equal quantity of matter exists both before and after the experiment; the quality and quantity of the elements remain precisely the same; and nothing takes place beyond changes and modifications in the combination of these elements.[67]

In the digital age, this principle is no longer true. After a power failure, the amount of unbuffered computational matter in a computer's RAM memory literally disappears without transforming itself. That's just hair-raising! Who has ever seen a material with the audacity to behave this way? What is this incredible phenomenology of disappearance about?

Let's take an even more commonplace example that everyone has experienced hundreds of times (even if it is technically more complex than it will appear here): I am quietly sitting at my screen, and I *delete* a file from my computer or my external hard drive. What happens? Where does it go? Is it transformed, or has it disappeared? Skeptics will say that it *has* changed because it has moved into the system's trash. For sure. But what if I empty the trash? What happens to it? Is it evacuated through some conduit? Where is the smoke? Where are the ashes? No need to look for traces. This time the file has really disappeared, at least at the phenomenal level that the user experiences through the interface (since it still exists on the hard drive). It did not change; it has not changed state. It simply no longer exists. We must take the measure of this event in all of its ontophanic, even ontological power: from the perspective of the world of experiences, what occurs is a nearly miraculous, instant slippage from being to nothingness! Without any other process. At most, the system made a tiny, friendly sound to confirm the operation. Has anyone ever seen such a thing? In the history of science and technology, has anyone ever heard of physical realities disappearing without leaving any visible trace?

Jean-Pierre Séris states that "there is more in technology itself than in all that philosophies have said about it to date."[68] That's even truer of digital technologies. Not only are there hidden phenomena in computers, which technology haters have missed, but also computers produce the most unprecedented ontological events, which put into question the oldest and surest principles of science—from a visible perspective. Imagine casting coke in a blast furnace without visible smoke! Imagine a nuclear power plant without atomic waste after nuclear fission! Impossible. In contrast, there is no need to strain to imagine that deleting ten gigabytes of digital data leaves nothing at all. Unplug your computer at the wrong time, and it will take the computer just a few seconds.

Computational matter is definitely a strange material. It introduces into the field of our experience an ontophany of disappearance to which we are slowly becoming accustomed. Rather than succumbing the dream of the unreal, we gradually begin to accept, even if it sometimes costs us, that some matter can vanish.

The digital phenomenon is self-destructive. This is the ninth characteristic of digital ontophany.

Fluidity: The Digital Phenomenon Is Thaumaturgical

Instant reproducibility, reversibility, self-destructibility: these three features allow us to see another one: digital phenomena are endowed with pseudosupernatural and pseudomiraculous powers. Indeed, everything we can do with computational matter seems easy and light, immediate and simple. With instant messaging, we have not only abolished distances, but also, and especially, weighty and arduous old procedures. No doubt this relates to technical progress in general, as the history of transportation technologies shows us: from stagecoach to bullet trains, we have gained in speed, comfort, and efficiency. But with digital phenomena, we have gained efficiency, and also flexibility, ease, and lightness. Not only are our emails sent and forwarded much faster than by postal mail, but it's a lot easier, more immediate, and simpler to write and send a message. Where once we had to make an effort—pull out a sheet of paper and a pen; find a flat and solid surface on which to rest; apply our hand to forming letters legibly; fold and put the letter in an envelope and with a stamp; find a mailbox; wait for its routing and delivery—everything is easier and instantaneous. No need to wait to get home or to the office. On the bus, on the street, at the beach: simply tap away furtively on your mobile keyboard and your message is simultaneously and instantly put in an envelope, stamped, placed in your outbox, and sent off, with one click. Yann Leroux says this:

> In the digital world, everything glides: I write, and letters appear one after another on my screen. I make a mistake, and I erase them without difficulty. Little effort is made to write, and the same goes for erasing. That is unique: remember our school books, the pleasure of the pen sliding across the page and of the eraser erasing without (almost) leaving any trace.[69]

This is what we are calling digital thaumaturgy, this pseudomiraculous phenomenology where things have lost their old gravitas to become instead light and fluid, that sweet phenomenality where things are airier and looser, magically bending to our expectations and our desires without the resistance of former times.

Philippe Quéau grasped this aspect of the digital phenomenon very early on when he became interested in virtual images: "Reality, precisely, is what resists against us. What the real world is is not up to us. ... What is the virtual? It is, it seems, quite the opposite. It does not resist, it becomes liquid, gaseous before our desires."[70]

Being in a gaseous state: that's what being online or living among interactions is in the age of digital interfaces. That's entering a new ontophany of procedures, in the sense that our experience of the world is the sum total of procedures—completed or yet-to-accomplish actions. Today, everything there is to accomplish is done through digital channels. Therefore, everything that can be accomplished is accomplished more lightly and fluidly—whether sending messages, booking airline tickets, buying music, paying bills, filing tax returns, editing photographs, publishing books. With our digital means, every procedure in life is simpler and easier.

This way of "living in a gaseous state" led us previously to considering that an interactive experiment was about being "detached from the body."[71] But to the extent that digital ontophany affects the body of things as a whole (that is, the phenomenological presence of things themselves), it is perhaps not exactly about being detached from the body. In all eras, whatever the technological system, we always have a body and never leave it, regardless of the technologies we are using. When we are online, our bodies are not absent. As Antonio Casilli has shown, "Communications on the internet ... swarm with 'body marks,'"[72] that is, with representations of the body, whether one-dimensional (usernames, smileys, pokes), or two-dimensional (2D avatars, photographs, biographies, and profiles), or three-dimensional (3D avatars, virtual characters): digital interfaces do not deprive us of our bodies, but they modify the ontophany, the way it appears. When I'm facing an interface, I'm not deprived of the experience of my own body, but I'm more focused on representations that I give it online. It is my psychic world that is primarily engaged by the interface—so much so that we have spoken of "psychic acceleration,"[73] but in so doing, my body is still participating in the phenomenology of the world that I experience by digital means, "if only thanks to the photons that come knocking at [my] retina,"[74] or, through the new game consoles with gestural interfaces using a whole range of body movements embodied in space.

Therefore, according to the phenomeno-technological lesson, it's not in the body but in the device that we have to find what is particular to the digital procedures of our world. And if there is any detachment, rather than being detached from the body, it's about being detached from the resistance of things, even though resistance is traditionally associated with bodies and materiality. Digital phenomena free us from significant parts of the capacity of reality to resist us. That's what digital thaumaturgy is about, this pseudomiraculous ontophany of fluid procedures, which makes of the world and of the experiences we have of that world something lighter and easier.

The digital phenomenon is indeed therefore like a thaumaturgic king: it performs miracles or, more simply, works wonders. That is the tenth characteristic of digital ontophany.

Ludogeneity: The Digital Phenomenon Is Playable

Video games are a total digital object. They bring together in one apparatus the art and technique of narration, graphics, moving images, music, but also interactivity, simulation, reversibility, or reticularity. After decades of evolution, it is safe to say that it is one of those objects that fully realize digital technology's potential. From the arcade games of the 1970s to living room consoles of 1980 to 1990[75] and to the online role-playing games of the 2000s, the spectacular history of video games is not only the story of the birth of the largest cultural industry in the world; it is also, and especially, that of the powerful rise of the ludic phenomenon. Video games increasingly export more and more of their code and culture, as much to other systems and sectors of the digital technological system as to more varied social practices. This is what is called gamification. It designates all "connected devices [that] enable game mechanics to be transposed to all of everyday life.[76] Although gamification can be seen in online consumption, communication, advertising, or vocational training, it is the subject of serious criticism insofar as it tends to bring into our lives the only formal game mechanics, such as scoring points ("pointification"[77]) without necessarily generating any fun, that is, any authentic ludic pleasure.

This is probably one of the reasons French researcher Sébastien Genvo chooses to talk about the "ludification"[78] of the digital to designate not the fashionable technologies of gamification but the fact that ever increasingly, digital devices stimulate our "playful attitude" even though they are not (and do not attempt to appear as) video games. Our times are more and more attracted to fun and enjoyment, either through playful forms of existence that mainly—and more and more—rely on digital devices that need not be video games. With ludification, "computer games today can not be subsumed under products from the video game industry": we are facing the "multiple forms *playful* takes."[79] Facebook is a good example of this. Here is a digital system where users have many opportunities to experience authentic fun but is not in itself a video game (even though it includes game-like applications).

In historical hindsight, one can wonder whether all of the efforts engineers and designers have made since the beginnings of microcomputing to make computers easier to use is one huge process of ludification to get computers from programmable machines to playable machines. It is more pleasant to click on buttons than to type out lines of code at the keyboard (at least for most people).

And from easing to pleasing, there is only a single step. With an interface, little is needed for a user to adopt a player's behavior in the sense that the playfully alliterative *pleasure of play* designates the pleasing activity of the game (as an experience), as opposed to the game itself, which references the formal system (as a set of rules and mechanisms).[80] Then, playing starts from the moment I interact with my environment for the sole purpose of deriving pleasure. To play is to *enjoy*. In a well-known passage of his *Aesthetics*, as Hegel wonders about the origins of the need for art, he illustrates this pleasure of interacting with the outside world which is the core of the fun (and artistic) activity:

> Even a child's first impulse involves this practical alteration of external things; a boy throws stones into the river and now marvels at the circles drawn in the water as an effect in which he gains an intuition of something that is his own doing. This need runs through the most diversiform phenomena up to that mode of self-production in external things which is present in the work of art.[81]

The same phenomenon occurs before any digital interface. Just as we admire circles in water, the simple act of seeing how an interface behaves when we interact with it is spontaneously playful. Who hasn't skimmed a drop-down menu on a website without any intention of clicking on any of the links, simply for the pleasure of seeing *what it does* or *what happens*? Will the submenu just drop down? Totally change colors? Gradually open up in an animated fade effect? Or trigger a complete redesign of the layout? In order to know, you have to try. And to try it is playful. The adoption of a playful attitude is therefore almost immediate and natural when looking at an interface. And it is true, as Bernard Darras points out, that "the prowess and performances of the 'mechanics' are so fascinating that even their random use is already a source of enchantment."[82]

That's why the digital is not only subject to ludification processes; it is inherently *ludogenic*, a term by which I designate the fact that it spontaneously promotes a playful attitude and stimulates our ability to play—hence, showing how successful computers are among children regardless of the kind of computer: workstations, consoles, tablets, smartphones, and others. This success is not only due to the power of the image to attract (which is also true for television), but also to the power of immersion in interactivity. No wonder then, since "the preferred and most intensive occupation of the child is to play games,"[83] that digital interfaces exert a particular attraction on children. They are fundamentally ludogenic. And it emerges that a "culture of cool" has permanently established itself on the Web and social networks: it's a culture of "chill," where humor and wit—other forms of play—are widespread and affect all users. Admittedly, digital interfaces can also inhibit some people. But if they were not fundamentally ludogenic, could they make cats so attentive as we see in a whole host of videos posted on the internet by people fascinated by the instinctive feline play of their cats with an iPad? Play is not a characteristic of humans only, but playability is characteristic of the digital.

Therefore, not only is "gameplay essential to the existence of any ludic phenomenon,"[84] but it is also an essential component of any digital phenomenon. Digital devices are ludogenic, in the phenomenological sense that they make our experience-of-the-world flow into gameplay. That is why we live in an ever more gamified or *playable* world: not only because, for commercial purposes, companies exploit the code of videogame culture, but also and especially because digital phenomena are ludogenic in themselves.

Such is the eleventh characteristic of digital ontophany.

6 The (Digital) Design of Experience

> Without us even realizing it, a new kind of human being was born in the brief period of time that separates us from the 1970s. He or she no longer has the same body or the same life expectancy. They no longer communicate in the same way; they no longer perceive the same world; they no longer live in the same Nature or inhabit the same space.
>
> —Michel Serres, *Thumbelina*

Since the end of the twentieth century, we have learned to live in contact with digital information. Its groundbreaking properties make our possible world experiences flow into new phenomenological prisms. Faced with the phenomenality of digital beings, what we learned to consider as real falls into disuse, forcing us to relearn to perceive in order to embrace a new ontophany of the world.

This long-term learning started around the 1970s and continues as we learn to *take things at interface value*. Besides, because we are experiencing an ontophanic transition, we are better placed than ever before to understand that the way things appear—ontophany—directly determines the nature of our experience of them. In other words, the quality of our experience of life depends on the devices that surround us—and how, as phenomeno-technological instruments, they make the world and give it to us. Under these circumstances, those who are tasked with designing these devices should be considered philosophically responsible for the experience, that is, for all that is offered up to us to be perceived, experienced, felt. Besides, digital ontophany is not only a new phenomenology; it involves a creative phenomenology that results from a process of fabrication.

The Fabrication of Ontophany

If all ontophany is a phenomeno-technological result, then each ontophany is a fabricated thing. For sure, in everyday life, architects, engineers, and designers do not have the feeling that they are fabricating ontophany. They work on spaces, objects, services, and interfaces. But indirectly, these spaces, objects, services, and interfaces are ontophanic operators. They structure our possible experiences-of-the-world. Therefore, architects, engineers, designers, and, more generally, all designers have a philosophical responsibility: that of being generators of ontophany or makers of being-in-the-world. Unbeknown to them or not, they decide on the phenomenality of phenomena, compose the ontophanic framework of our existence, and, because they are working on user experience, choose which possible world experiences are available.

To exist or to be-in-the-world implies that the world has a form that can make us be. That's why Peter Sloterdijk defines it as a "sphere." By that, he means a world that is shaped: "A sphere is a world formatted by its inhabitants."[1] From then on, to be-in-the-world is to be-in-a-sphere in the sense that a sphere, on the model of the Greek idea of a house, implies a "reciprocal belonging" between the place and its inhabitants.[2] In other words, the sphere is the "self-centered world" (Jakob von Uexküll's *Umwelt*) that we create in our (technological) effort to format the "existential." Commenting on Sloterdijk, Bruno Latour explains it as follows:

> To try to philosophize about what it is to be *thrown into the world* without defining more precisely, more literally (Sloterdijk is first of all a literalist in his use of metaphors) the sort of envelopes into which humans are thrown, would be like trying to kick a cosmonaut into outer space without a spacesuit. Naked humans are as rare as naked cosmonauts. To define humans is to define the envelopes, the life support systems, the *Umwelt* that make it possible for them to breathe.[3]

These envelopes or life support systems are phenomenologically what we have called phenomeno-technological devices or ontophanic devices: they act as techno-transcendental structures that shape our way of perceiving and cast our being-in-the-world. Therefore, all of those whose job is to design and implement form have a major responsibility. They contribute directly to the creation of our existential sphere, conceived as the *Umwelt* in which we exist phenomenologically.

For example, in designing the National Library of France as a monumental palace with abstract and cold exterior lines while offering a safe retreat on the inside that escapes from the city around a forest pine, French architect Dominique Perrault has made of the reading rooms at the François-Mitterrand site a haven for peace of mind, which promotes concentration. We could never have experienced the same soothing intellectual experience without the phenomenological quality of this architecture and its furniture. Nevertheless, by denying to the library, for aesthetic reasons, the right to affix blinds at the large bay windows of its interior facades, the architect inflicted on his building a technical inability to protect its users during long hours of sunshine that can temporarily make parts of the reading rooms unusable—except for accepting to sit through the discomfort. This double-edged example shows that it takes little for our experience of the world to be wonderful or horrible depending on whether the technology that shapes it was more or less well designed—proof that designers are the true shapers of the world we live in, whether in libraries, on the street, in housing, in transportation, in the countryside, or with ubiquitous digital interfaces.

There emerge several important consequences. The first is that technology cannot be anything but a practice of shaping our existential sphere and, as such, cannot be dissociated from the activities of design creation. All that exists in our sphere is a technological result of creative and design operations, hence an interest in speaking of "material culture" and giving up the purifying distinctions between art and the technique. The second is that if experience is a phenomeno-technological construction, its construction process is precisely one that supports these creative and design operations. Nothing could indeed be built (phenomenologically) without (physical) construction operators. In order for experience, that is, the very fact of perception, to be a (techno-transcendental) co-construction of meaning and devices, there must be some shared constructivity between our perceptual abilities and the creative operations of our technological devices. Both necessarily play an active role in the constructivist process of developing experience, and that is why any ontophany is fabricated. Therefore, since they are fundamentally practices that give shape to our sphere, design-creation activities play a role major in the ontophanic quality of our experience of the world. We propose defining them not only as technological activities but also as phenomeno-technological activities. In this perspective, all technological or artistic acts, all fabrication of artifacts, all the shaping practices emerging from the material culture are phenomeno-technological activities that lead our possible experience-of-the-world to flow into ontophanic regimes proper. So is it with painting, literature, film, and video games, "with their own pleasures, with their very own regimes of experience, with their particular sensitivities,"[4] but also with craft, engineering, or design. All of this affects the ontophany of the world and, consequently, our possible experience, by playing a role in the construction of our existential sphere. Among them, we still need to pay special attention to design, whose phenomeno-technological capabilities are unique and edifying.

Design and Factitive Intentionality

French semiotician Anne Beyaert-Geslin offers some reflections on objects in a 2009 article in *MEI: Mediation Et Information*: "Forms of Tables, Forms of Life."[5] Rather than thinking about the object in the aesthetic terms of form-function (as did Louis Sullivan)[6] or in sociological form-sign terms (as did Jean Baudrillard),[7] she approaches it in the semiotic terms of form-action. By borrowing this concept from Jacques Fontanille (who got it from Algirdas Julien Greimas), she defines an object as a "factitive" object, that is, an object "that has [someone/something] do, be, or believe."[8]

To do this, Anne Beyaert-Geslin compares the shape of a table from the Middle Ages with that of an eighteenth-century table. On the long side of a medieval table, its valued side, she explains, there is a bench where the master of the house takes his seat in the middle, most often alone, with his back at the fire, while stools are arranged along the width, making face-to-face encounters difficult. Unsettled, this medieval table is not fixed; it is mobile, made of planks on trestles, and erected for each occasion; utensils are shared among guests. But dining rooms make their appearance in the eighteenth century, "with richly decorated rooms where preferences are soon for round tables" and where the appearance of place settings recreates a personal sphere around the plate. Through these examples, we understand how the specific shape of a table generates a particular form of life.

In both cases, in experiencing the meal, the relationship to the other is conditioned by the object:

> In the Middle Ages, the table is mobile and off center; in the eighteenth century, it becomes fixed and stands out as a central marker of a room whose purpose it determines. The earlier scene marked by individualism and nomadism, structured around a single actor it accompanies through space, transforms itself from then

on into a collective and sedentary scene that invites the actors to take part in a common activity located in one place once and for all.⁹

Such is the *factitivity* of the table: placed once again in its practical scene, the table both *makes* the meal *be* in a way that the table conditions and gives the guests an experience of being together that the table structures a priori. Once again we encounter the phenomeno-technological idea in all its strength: artifacts are apparatuses that, acting as techno-transcendental structures, *make* the world *be* (factitivity of make-be), as much as they condition the possible experience that we can *make* of it (factitivity of make-make). The phenomeno-technology of objects is therefore a form of phenomenological factitivity, for the objects technically build the scheme of possible experience that they make accessible. And what is true of the table is just as true of a chair:

> A chair is rarely isolated and takes its place at the meal table unless lined up with others; it allows us to wait our turn at the doctor, or, if we take care to arrange it with others in a small circle, for residents of a retirement home to have a conversation.¹⁰

Objects make the world. In this sense, all artifacts are factive, and phenomeno-technological activities are those that can be identified by this capacity to *make-be* and *make-make*.

Among such activities, design occupies a unique and, to put it all out there, exceptional position. Indeed, objects that are not the result of a design process are factive only by accident; the world in which we live flows despite them because they are phenomeno-technological by nature, like all artifacts. By contrast, objects resulting from a design process are recognized as being intentionally factive. They are designed to *make-be* and to *make-make*, that is, to generate new ontophanies and reshape possible experience, with a view to "improving or at least maintaining the habitability of the world,"¹¹ according to Alain Findeli's beautiful turn of phrase. Design factitivity is therefore not an accidental feature of its phenomeno-technological nature; it is an intentional feature of its creative culture.

That's why designers are neither artists nor engineers. Designers are "projectors."¹² One needs to understand this as meaning that in the most basic sense as "projectors" of ontophany. Designers always intentionally seek to produce an "experience effect,"¹³ that is, an "ontophanic effect," to transform raw, nonquality uses into highly qualitative "experiences-to-be-lived."¹⁴

This factive intentionality, which directly aims at increasing the quality of lived experience, is the reason design is not a phenomeno-technological activity like any other. Design is not just about building an experience, as are all technical activities. It is meant to delight. This is the meaning of this intention (to enhance experience) of *make-being* and *make-making*—design's intention. Besides, we ought not confuse the field of design with the field of objects, which—tech haters have told us enough—are far from all being bearers of delight (poorly designed devices, unsuitable architectures, harmful artifacts, destructive technologies). Better to consider it a field of effects, in the phenomenological sense we give this term,[15] that is, these "design effects"[16] that happen in objects, spaces, services, or interfaces to improve our existential spheres.[17]

In other words, it is not because one manufactures an object that one is designing. To practice design, one must be able to *project* into an object some factive delight. When Anne Beyaert-Geslin cites as an example the Joyn table of the Ronan and Erwan Bouroullec brothers, she is describing an entirely different factivity than the tables of the Middle Ages or the eighteenth century. She is describing an intentional factivity. Why? Because a Joyn table, in the mind of its creators, is precisely designed to be "a collective desk where one works on a laptop while others eat or talk."[18] It is modeled after kitchen tables of the past.[19] That a Joyn table suggests doing otherwise—to "reconcile office work with conviviality, a part of life that seems *a priori* contradictory"[20]—it is because its designers projected into it this factive imagination. Further proof of this intentionality is that the whole world is not ready to make full use of it, judging from the reductive use that is often made of this table, reduced as it often is to the status of an "ordinary collective office desk where everyone can avoid the vagaries of promiscuity and retrench to an area."[21]

Design is therefore a phenomeno-technological activity that is *intentionally* factive. As such, it looks for new ways to appear (*make-be*) in an effort to create other possible experiences of the world (*make-make*). At its best, design is an activity that creates ontophany. It has a fundamental ability to make worlds or, as Peter Sloterdijk would say, to create a sphere. In short, it is an intentionally factive activity that seeks to introduce delight into our sphere of existence. In the age of the digital technological system, which offers new material to work with, the role of design is essential to the constitution of a new ontophany and the quality of possible existence of the regimes of experience that it carries.

Digital Design Effects and What They Make Possible

Today digital design is the most innovative type of design. It can be defined as a creative activity of digital ontophany inasmuch as it makes up our new perceptual environment or *world proper* (*Umwelt*). Some of the fundamental properties of computational matter, such as the Dionysian virtuality of graphic interfaces, would not exist without the creative spark of the design gesture that is at its roots.

But to really understand what digital design is, we need to lay out a distinction at the outset.[22] There are two ways to think about the relationship of the digital to design. First, there is what we have called *digitally assisted design*; it covers design practices that exploit digital material as a simple medium. The second falls under digital design proper; it includes all design practices with factitive intentionality that make use of digital material as both a medium and an end. Let us examine each in detail.

The first uses computational material as a creative instrument whose purpose is to give life to uses (factitivity) by shaping materials with the help of a computer, including computational materials. Whether these uses play a part in our primary sensory experience (formatting physicomechanical materials for unmediated perception) or in interactive virtual environments (formatting electrodigital materials with *interfaced* perception) does not make any difference. In both cases, digital material is used as an instrument just like a pencil and a square, or as a method like gluing or folding. For example, designing a bicycle helmet using design software, making a vase using stereo-lithography techniques, printing a 3D object: these are design practices that fall under digitally assisted design. The process is digital, but the product is not.

The second resorts to computational matter as matter to be molded in and for itself intending to give life (factivity) to computational materials. That does mean that computational matter is itself the end purpose of a design process, which would make no sense since design does not aim to produce materials but rather to promote regimes for better experiences. More to the point, then, computational matter is itself part of the factitive intentionality of the process; it is intentionally projected into the project as a necessary material component of the final result, in light of its novel potential. It is not the end purpose of the project, but it is indeed part of its factitive goal. We can thus speak of digital factitive intentionality in the sense that design gestures are rooted in an intention to *make-be* and *make-make* that makes preferential use of digital material and its new (and exceptional) ontophanic properties.

In concrete terms, this means that digital designers are the ones who, knowingly or not, engage in a project whose results will be made of computational matter. For example, designing a touch pad or networked object, developing software or creating a mobile interface, producing a website or designing a video: all this and much more is about digital design because the process has to be digital and the product too.

And that changes everything. Unknown ontophanic properties of computational matter offer designers the possibility of imagining experiential arrangements that are themselves unknown, such as those that may have been born of the virtuality of graphical interfaces since the 1980s. The task of digital design is therefore to creatively make use of the ontophanic capabilities of computational matter to create new design effects, that is, effects of the factitive delight of experience. The case of game design, as a generator of regimes for experiencing video games, is exemplary in this regard. To fully grasp the digital design effect operating in game design, one has to understand the unique nature of the ludic video experience. In his *Philosophie des jeux vidéo* Mathieu Triclot describes it as a form of "instrumented experience" that follows a long history of similar experiences:

> Culture has always been a matter of technology. We use technical devices or more or less elaborate artifacts—books, movies, movie theaters or theaters, concerts, paint canvases, etc.—to produce or rather to support the production of certain forms of experience. On the other side of the device, its workings, its technical possibilities, and its architectures, there are these little states, bookish, filmic, or fun [ludic], that are to be produced, that we care for and love, with their own pleasures, their very own regimes of experience, and their particular sensitivity.[23]

We knew this: experience is a phenomeno-technological construction. Mathieu Triclot emphasizes that artistic and cultural experiences are no exception to this. They are all made using instruments, devices. No doubt it was the phenomenological violence of the digital revolution that brought this to light for us finally—an effort of the gaze that no philosophy of technology could ignore any longer:

> For the technical nature of video games or movies to strike us today more than the book's is only possible because we have forgotten all the constraints of the book-as-object, all the rigors of writing or training that "graphic reasoning" implies for thought. There needs to be a whole effort of the gaze to make the technical nature of writing and the book reemerge, the one we make use of precisely when reading.[24]

This effort of the gaze must be ours. Given the variety of video game devices (terminals, desktops, consoles, mobile devices), "playing is never anything but enjoying these devices to generate experiences, to get into a certain state."[25] But to get into this state, you need a game with enough gameplay. And gameplay generally amounts to all the game mechanics that generate authentic, playful pleasure. That's why only a *game design* gesture—maybe we should call it *gameplay design*—can fully achieve the authentic experience regime of play characteristic of video games—in a gesture of *game design*, as in any other gesture of digital design, a factitive intentionality, that is, a project of ontophanic delight of the experience (of play, here). On this point, the history of video game technologies confirms the history of all phenomeno-technology: "Each time, what is invented are new connections to the machine, new regimes of experience, new ways to enjoy the screen."[26]

The video game regime of experience, as an unprecedented regime of experience, is therefore the regime of a particular screen pleasure that Mathieu Triclot depicts as "a specific and gigantic pleasure."[27] Only video game technology can engender this specific pleasure and, perhaps, only a gamer can understand it, because it derives directly from the exceptional properties of digital ontophany, such as interactivity or reversibility, as we have seen.

To produce unknown regimes of experience is the role of digital design. And what is true for game design is just as true for interaction design (computerized objects, networked things) or web design (websites, applications, online services). In the case of web design, for example, the designer's choices are never without consequences on the regime of users' possible experiences. As Bernard Darras has shown, the choice of layout, colors, or shapes that make up the home page of a website is not about mere aesthetic or personal preferences; this choice "determines the experience of the user by locking his/her interpretative process" because of "the halo effect" produced by the "first visual impression" in the first tenth of a second visiting a site.[28] That's why we sometimes want to leave a website immediately, whereas other sites, to the contrary, stimulate our pleasure to browse and explore them.

To all who wonder what to do with the digital revolution, it is easy to answer in a word: you have to design it. To be honest, this has been the case for thirty years, but we need more than ever before to continue to shape digital ontophany in a way that makes sense to us humans.

Interactive Situations and Our Ontophanic Future

Our interactive situations today a priori condition most of our existential experiences; that means the form that the experience-of-the-world takes when it is ontophanically generated by a digital interface. Today, existential experiences for the most part line up with interactive situations, whether in our professional or personal, public or private lives. They are also very varied. They range from interactive desktop situations, modeled on workstations, to the many forms that rely on some sort of apparatus: interactive situations brought about by a website, a mobile terminal, a tablet, a video game, an e-book, a networked thing, or an interactive spatial device, among others. It would be laborious indeed to try to enumerate all possible forms of interactive situations. These forms, long reduced to personal computer workstations, are increasingly numerous, innovative, and unexpected, given that computational matter is currently seeking a home in so many of our human artifacts, smart cities, self-driving cars, or humanoid robots.[29]

In addition, the current human condition is leaning heavily toward becoming a generalized interactive situation. Contemporary humans are primarily interactive beings who are constantly manipulating digital interfaces at home, at work, in transportation, on the street, privately. Shaped by the novel, ontophanic properties of computational matter, their experiences of the world are a form of life where one finds the contemporary existential trait of immersion more and more present, though to differing degrees. As Sherry Turkle reminds us, immersion is precisely what the digital interfaces of our virtual environments require of us:

> In a design seminar, the master architect Louis I. Kahn once famously asked: "What does a brick want?" It was the right question to open a discussion on the built environment. Here, I borrow the spirit of this question to ask: "What does

simulation want?" On one level, the answer to this second question is simple: simulations want, even demand, immersion.[30]

We all experience this. Interfaces are attention grabbers, not only because they offer us captivating, simulated environments (virtuality), but because they have this set of novel properties that have the exceptional power to motivate us (interactivity, reversibility, reticularity, fluidity, ludogeneity). There is no need to be in a persistent, virtual world. In order to learn about immersion, all one needs to do is interact with a smartphone every day, or a tablet, or a microcomputer. In the era of the digital technological system, immersion is everyone's ontophanic condition, and it is becoming more and more commonplace. Those who love this way of being-in-the-world call themselves geeks. Those who like it the least willingly criticize it on the grounds of a loss of reality or authenticity. This bipolarity has nevertheless existed since we started introducing interfaces into our daily lives.

As Sherry Turkle shows, as soon as the first computers make their way into professional practices, at MIT, for example, in the mid-1980s, there are two types of reactions: on the one hand, the enthusiastic adoption of those who yield to the immersion of enticing machines and start to create using computers (these are the *makers*); on the other hand, the worried skepticism of those who express great distrust of these new tools and who doubt their relevance, fearing a loss of reality (these are the *skeptics*). The same enthusiasm exists among supporters of social media, and the same skepticism is at work among those who argue for days without screens or web detox, such as those called "disconnectionists.[31]

But we now have to recognize that the dialectic between enthusiasm and skepticism is not about two opposing sides. In fact, it exists within each of us, we who are living out the transition between two ontophanic ages. Sometimes we immerse ourselves in interfaces with pleasure; sometimes we feel enslaved by them. Sometimes digital is good, sometimes bad. Why is that? Because that is generally the cost when adopting a new ontophanic culture. One does not cross the valley that leads from the old world to the new without meandering. But especially, ontophanic cultures accumulate as much as they succeed each other, so that we can move from one to the next without moving through to the other side of the mirror. Digital ontophany has not made telephonic ontophany or face-to-face ontophany disappear. It has simply relocated each one to reflect to the potentialities we are interested in using. For example, under the effect of *digital relationships*, face-to-face ontophany acquires a greater phenomenological aura, which we reserve for certain people or situations—whereas in the past we had no choice but to engage face to face. Similarly, telephonic ontophany acquired a phenomenological aura that we had not encountered earlier; the telepresence of the human voice, even restored by a device, has such perceptual force that we would rather save telephone interactions for certain people or situations.

Then, we keep more reticular ontophanic modalities for those of our relationships that sociologists might call "weak ties" (such as so-called friends on Facebook), the number of which has increased considerably. The modalities involve written messages sent over the internet without great phenomenological aura because they evaporate as fast as they are routed—even if they do have a strong emotional charge and extraordinarily real effect. Various existing technologies make it possible for us to choose the distance that suits the relationship we wish to have with others.

Despite its omnipresence, the immersive experience of the world is never just a form of experience that we add to others. It becomes commonplace, and we gradually end up taming and locating it in our hybrid ontophanic sphere because we live at a time when ontophanic culture has never been as rich and diverse. Modern humans are a bit like navigators of their ontophanic ships. Depending on the winds and the tides they encounter, they decide which regime they want to give their existential experiences to, at times turning portside to the side of digital ontophany or veering starboard toward face-to-face ontophany. Between the two, they have a full range of ontophanic tones, from telephones to messaging, by way of email and social media. It is up to each person to evaluate where at any moment to place the phenomenological cursor of his or her existence and relationships with others, according to his or her interlocutor, mood, and so on.

The future of our being-in-the-world has never before been so connected to design, an ontophanic, delighting, creative activity. In times of a generalized interactive situation, it is incumbent on digital design to sculpt our possible experience-of-the-world by exploring the various ways of making immersion a factor of delight. This does not mean just producing quality interactive experiences that give meaning to immersion itself, which is already remarkable. It also means seeking to combine the ontophany of immersion with other ontophanic cultures and seeking to produce hybrid regimes of experience that use all the ontophanic facets of the real and use the best of each. Immersion on its own cannot be an exclusive end in itself. It is interesting only to the extent that it enriches our possible global experience-of-the-world.

The question is not, then, whether to abandon printed books in favor of electronic books or board games in favor of video games; whether to give up learning to write on paper in favor of learning to write on a tablet; or whether to transform all our face-to-face relationships into digital connections. Living exclusively in an immersive state, in a restrictive digital ontophany, can only be a phenomenological impoverishment of the experience of existing. The question is not either whether to do exactly the opposite: stand against e-books, forbid video games, reject tablets in school, or close your Twitter account, waiting for the grass to grow. As attested to by Paul Miller, who spent a year disconnected from all of his screens, only to discover that he had made a mistake,[32] the experience of living off the digital grid will never lead to finding some miraculous, *truer* (because it is more *natural*) original, ontological glow. We simply have never existed in such a state of nature. Any ontophany in the world is a techno-ontophany. When one forces oneself to live outside the dominant, ontophanic culture, one finds only an other, older one, anchored in other devices. As Jean-Claude Beaune writes, "The world we face, in our most mundane experiences, is cultural; that is, technological and technicized through and through. We have no natural experience of the world or of ourselves."[33]

Rather than categorically being against digital ontophany or succumbing to it blindly, it seems appropriate to try to use the best phenomenotechnological capabilities of each techno-ontophany and its hybridization with others. Hence our interest in a device like Apple's iPad, which frees us from our workstations; a console like the Nintendo Wii, which boosts physical activity; games such as Éditions Volumiques' video toys,[34] which put screens in things and things in screens. That is the responsibility of digital design: to get us to live better lives in our digital, ontophanic environment, conceived of as a perceptual environment of digitally centered phenomenality, but a hybrid one, both digital and nondigital, both online and offline. This is the environment in which our children are growing up; it is where they are assimilating new structures of perception and acquiring the meaning of reality—their own reality.

Conclusion: On the Radical Aura of Things

If I had a hammer
I'd hammer in the morning
I'd hammer in the evening
All over this land
I'd hammer out danger
I'd hammer out a warning
I'd hammer out love between
My brothers and my sisters,
All over this land
—Pete Seeger and Lee Hayes, "If I Had a Hammer" (1949)

In 1992, in *Aramis, or the Love of Technology*, Bruno Latour said he wanted to "convince [us] that the machines by which [we] are surrounded are cultural objects worthy of [our] attention and respect."[1] In his own way, he was following Gilbert Simondon's call thirty years earlier to grant technological objects a place in the world of meanings. He attested to the persistent resistance of contemporary thinkers to take seriously the "technological dimension of acts of culture."[2] Twenty years later, one can say that under the effect of the digital revolution, no one can any longer escape the awareness of the meaning of technological objects and, more generally, the need to no longer think of technologies as objects that are separate from subjects. Our being-in-the-world is itself a fact that is technologically produced, and our ability to perceive, formerly one with natural procedure, depends on the apparatuses of the technological system in which we live. The digital revolution, far from being just a technology revolution among objects, is first and foremost a phenomenological revolution in subjects; it produces a new phenomeno-technological casting of the world and reshapes the ability of human beings to experience it.

It is therefore no longer possible to surrender to the humanist illusion. If, as Bruno Latour has it, "we have never been modern,"[3] it is indeed because so far we have never stopped being humanists. And "humanists only feel concerned by humans; the rest, for them, is only pure materiality or cold objectivity."[4] That is why the digital revolution works like a digital revelation: by having us discover the technological dimension of the question of being, it finally makes us become modern, that is, humanists and technologists at the same time, far from any easy transhumanism or extravagant posthumanism. Man is in the machine as much as the machine is in man. And we must add to the phenomenology of intersubjectivity that of interobjectivity. We are among objects as much as among subjects. To live is not only to live with "my brothers and my sisters"; it also means to live with "a hammer and a bell." It means living live with tables, chairs, shoes, cars, refrigerators, televisions, computers, architectures, landscapes—because landscapes are also technologies.[5] It is living with nonhumans, with whom we build ties all the time, to a greater extent than with other human beings (thinking here about the actor-network theory, so dear to Bruno Latour). Because unlike other human beings, of whom we can be deprived when we are alone, we are never separated from objects. Our ties with objects are permanent, which is why they can be lasting and deep and sometimes even become more important than some of our human relationships—with objects laden with memories or with a childhood landscape, for example. But what proves even more the intensity of our being-in-the-world-with-objects is that we put as much effort into selecting the things we want to live with as the people with whom we wish to maintain relationships, as evidenced by the success and general fashion of interior design. For Peter Sloterdijk, it is even a feature of the period:

> We are now living an epoch in which a more or less satisfied and luxurious conscience is learning the art of arranging its space. Modern man is a sort of "curator" ... which is to say, an exhibition planner of the space that he himself inhabits. Every man has become a museum curator. ... After the actual destruction of so many things and proof of the destructibility of everything, each inhabitant, in no matter which apartment, city, or country, has become or been forced to become a kind of planner of his own place.[6]

Conclusion

In other words, we are all designers of our spheres of existence. We take special care to *design* our spaces, our kitchens, our living rooms, and to decorate them with objects supposed to express our taste and our deepest personality—as if it were impossible to be well-in-the-world without being-well-surrounded-by-finely-chosen-objects. We are at the point where our interiors look more and more like standardized showrooms and where we seem to believe that it is absolutely necessary to afford a Charles Eames chair so that the fact of sitting has meaning.

In truth, all of this only reveals our need for objects and shows that we cannot contemplate existing without them, since it is they that make our existence.

Not only do we bond with them, but also we phenomenologically build our experience-of-the-world thanks to them and through them. From this point of view, the existential quality of our possible world experience does not rest solely on the quality of our relationships to others but also on that of our ties to artifacts and on our ability to receive and appreciate their own phenomenological aura.

By *phenomenological aura* we must understand something close to the Benjaminian aura, that is, the *uniqueness of apparition* with which things are offered up to our perception. But we add to it the notion of degree. The phenomenological aura of a thing—whether an object or a subject—is its degree of perceptual intensity, phenomenal vivacity, ontophanic acuteness, and potential to appear. In that sense, all things are not equal because they don't all have the same phenomenological aura. Some have more, others less. This has nothing to do with their degree of reality. By definition, a being of any sort is endowed with existence, and therefore with reality, and this is the case of digital beings as of others. But not all beings have the same phenomenological aura; they do not all emit the same perceptual intensity. And this phenomenological aura is conditioned by apparatuses that cast their appearance phenomeno-technically. Some ontophanic matrices give off more or less a phenomenological aura. Thus, digital devices do not give off the same phenomenological aura that mechanical devices do. When I experience someone via telephonic ontophany, that someone has for me much more phenomenological aura than when I experience him or her via digital ontophany. But let's not make any mistakes: in both cases, they are just as real. One must distinguish the degree of existence of a thing—as a *quantum being*—from its degree of phenomenological aura—as *quantum perception*. A thing may have less of an aura than it has existence, and vice versa. And that is where the whole subtleness of the digital revolution as an ontophanic revolution lies, and that is what, for a long time, was deceiving us by throwing us into an illusion of the virtual and the dream of the unreal.

If the digital revolution is indeed a phenomenological revolution, it is precisely because it casts phenomena in an unprecedented phenomenological aura.

And—epitome of subtleness—this new phenomenal casting is paradoxical in that it begets things that, on one hand, have a weak phenomenological aura (online chat, for example, or friendships on Facebook) and, on the other, strong reality effects (strengthening social bonds through an online community or using Twitter to message more broadly). This is why we call our social media ties *weak*. On a philosophical level, they do bear some phenomenological weakness, even though they may bear great reality or sociability. The best proof of this is that when we are close to and intimate with someone, let alone in a relationship involving our bodies, the phenomenological potential of face-to-face ontophany crushes every other. In a couple's life these days, for example, one texts or messages just as readily as one shares things on Facebook. This obviously is no less real than when one is face-to-face, especially when it is about sending each other photos of your children. But because it happens online, through digital interfaces, it has less phenomenological aura—to the point where sometimes what is messaged by text or over the network is canceled out by what is exchanged face-to-face, which is the space of ontophanic restoration par excellence, the irreplaceable place of phenomenological authenticity, where misunderstandings sometimes produced in online exchanges are aired and rectified.

Face-to-face ontophany is therefore by far the one that has the highest degree of phenomenological aura. There are forms of existential experience that are indissociable from it and that, for that same reason, are incommensurate with digital ontophany, such as the experience of maternal tenderness, intimate love, or psychoanalytic treatment for that matter. With the help of the digital, which reveals, we understand how a psychoanalytic experience is produced phenomeno-technologically using such *technologies* as the couch, the armchair, and the four walls of an office. A psychoanalyst's office is an apparatus that grounds the potential appearance of the unconscious and insight. And at a time when the rise of digital devices makes alleged *remote psychoanalytic therapy* possible (for example, via Skype), psychoanalysts have to remind us of that fact, in an indispensable effort to define the specificity of their clinical practice compared to "skypanalysis." For Geneviève Lombard, a pioneer on this issue in France, the authenticity of the psychoanalytic experience requires "the effective presence of two people in the same place at the same time" because

> copresence via intervening screens, if it is really *real* and allows for real exchanges, where emotions and many aspects of human life have their place, does not yield the essence of what constitutes an authentic presence (body and soul) of one person to an other.[7]

Such an "authentic presence" can occur only with the irreplaceable, phenomenological aura of face-to-face ontophany. Although it proposes a miraculous phenomenology of the experience of existing, whose amazing and spectacular properties we have described, digital ontophany has limits, which are just as necessary to affirm as they are difficult to express (except in a caricature). The induced experiences of the world, which digital ontophany induces, despite the unprecedented power of their reality effects, has a rather low degree of phenomenological aura. That is why, before them, experiences of the world induced by face-to-face ontophany have never before been so important. And that goes not only for our face-to-face with subjects but objects too, because to exist or to-be-in-the-world is also to know how to appreciate the nonhuman moments, moments of encountering facts, opportunities of walking among things. To live is also to live with things and to know how to taste their phenomenological aura. Walter Benjamin always refers to "the aura of these mountains, of this branch," for example. He seeks things whose phenomenological aura is greater than that of photographs. In the same manner, experiences today persist whose phenomenological aura is greater than that of interfaces.

Supplement 1 Otherphany and Otherness

Revised excerpt of a text published in 2014 in the journal *Hermès*

The Other Is a Technologically Produced Phenomenon

The experience of otherness excellently reveals ontophany. By "the experience of otherness," we mean that this is not so much about otherness "in itself," in its radical essence, as much as otherness "for oneself," as it were, that is, as it gives itself over to consciousness in lived experience. We are investigating otherness as it is perceived, in and through historically time-stamped artifacts that surround the subject. How do others become manifest to me in the field of my experience? Our central thesis is that others are never separable from the technological conditions that make their appearances available to my perception, that is, to embody a presence that I experience.

The presence of the other, in its own phenomenality, is by no means natural; it is always technologically produced. This is what we are calling *otherphany*, that is, the way the other appears to us and gives itself over to us in our lived experience according to technical criteria. What seems at first sight incommensurate with technology, namely, the very fact of being other, in fact bears a techno-transcendentality, at least from the point of view of the phenomenological experience of the subject. Otherphany is therefore not otherness, only its phenomenological dimension.

S. Vial, "Ce que le numérique change à autrui: Introduction à la fabrique phénoménotechnique de l'altérité," *Hermès, La Revue* 68, no. 1 (2014): 151–157.

It designates the ontophany of an other, which is to say it results from the phenomenological production of otherness.

Three situations may suffice, at the phenomenological level, to demonstrate this.

Situation 1—In-person Otherphany

In-person phenomenality with others depends entirely on the technical/technological factors that structure its perceptual properties. Take the example of a conversation between two friends in a bar: the table and the chairs determine the appropriate distance (neither too far nor too close); the nature of the place favors the withdrawal from the world of work that is necessary for a free and pleasant exchange of words (a bar is a place of relaxation); the social environment determines the friendliness of the atmosphere (the people around you, who are also having a drink); and so on. The entire set of furnishings and architectural artifacts that make up this scene conditions the particular perceptual quality with which I, in this context, am given to perceive, listen, live, experience the other.

Situation 2—Telephonic Otherphany

In a telephone conversation, the way in which I sense others in the telephone medium is completely different again. The absence of body language and inability to see a face, the instantaneous and carnal presence of a voice and its intonations, are its immediate and central technoperceptual properties. It is about "talking to each other without seeing each other." The way others are given to me in telephonic ontophany and who therefore exist for me depends on the unique perceptual characteristics that stem directly from the tech properties of the telephone device, which clearly appears here in its very phenomeno-technological dimension. Just as in Bachelard, a particle accelerator generates the phenomenal reality of elementary particles, the telephone literally engenders the Other, at least as a phenomenon given to the field of my experience.

Situation 3—Loving Otherphany

To wrap up this persuading, let's deliberately use as one last example among our existential experiences of otherness, an otherness that seems at first blush the least technical/technological in the world: love. Although this is not usually the angle of approach for analyzing it and even though this allows us to get at only one aspect of it, we can say that the experience of love is also technicized or phenomeno-technologically produced. If love exists supremely in that instant when one "makes" it, then one has to take very seriously into consideration that there are always technoconditions that determine this *feasibility* and, not the least of which is precisely the characteristic of the possibility of this feasibility. To make love, you have to (most of time) hide, that is, escape the gaze of others by using technical/technological means (for example, four walls). To make love, you have to (most of time) use objects to support bodies (bed, couch, armchair, rug, table, chair, or whatever else you want); to indulge in some sorts of erotic play (sex toys, various accessories); or to guarantee comfort and safety (condom, pill, and so on). Generally we do not bother to think about it all, except precisely when we run out of something or something breaks down that makes lovemaking fail. In other words, experiencing love, this intimate experience of the self and the other, so often wrapped in an aura of romance and subjectivity, is actually an experience that, like all the others, is also realized and technicized, materialistic and objectivist. We are not asserting that this is its primary and essential dimension, only that this dimension exists just as much as the others for the experience to take *place* and release all its effects—proof that nothing happens in life experience without phenomeno-technological mediation.

Through these situations, it is our goal to show that the phenomenalized experience of otherness or otherphany is always a technologically conditioned experience. Whatever its complexity (in itself), an other is always given to me via phenomeno-technological mediation as a phenomenon (for itself) of my experience. In other words, in every manifestation of otherness, there is technicity or, more exactly, transcendental technicity. Transcendental technicity is the a priori technicity that structures the subject's possible experience by infusing the subject with an artifactual reference environment. Today ours is digital.

Toward Digital Otherphany: The Example of Online Friendship

For the past few years, digital social networks have offered a new artifactual environment in which a new otherphany takes shape, which can be called digital otherphany. This one concerns the particular way others appear to us through the digital medium, which is to say through computerized devices that organize our everyday lives. Such a manifestation of the other on the internet is at this point unprecedented phenomenologically; for a long time, it has seemed supernatural, as attested to early on by the use of the term *avatar* (which in its Hindu origins denotes divine incarnation) to point to the representation of a user on the internet or in a video game.

It is true that before the digital age, it was never possible to experience this: to *communicate* without speaking and seeing each other as one can in various transmedia texting practices such as messaging, tweeting, and chatting on Facebook. Digital otherphany, that is, the way in which others manifest today in the field of perceptual experience through digital artifacts, is therefore radically new. It is made of this paradoxical ambivalence that mixes presence and absence (on Skype, for example, the other is there without being there) and introduces a new way of experiencing the presence of others. And more and more, we become accustomed to it. For a long time now, we have accepted the otherphany of email, this new way of being connected to each other, which, "for many of us, [makes] cyberspace now part of the routines of everyday life."[1] In recent years, we have also learned to interact daily on social networks with our friends, our contacts, our circles, and as it stands, we are naturalizing this new form of social bond. By the same token, dating sites are becoming more and more legitimately accepted places for meeting people, for those who want to embrace them, as well as escape from the tired constraints and artifices of traditional social ways of making a match. Even remote online therapy sessions are becoming commonplace, and more and more people consult their therapists from their homes, using the internet despite the data confidentiality issues it creates or the problem of transplanting the framework of the psychotherapeutic setting.

Regardless, these various practices demonstrate that the relation to the other in its most varied forms is already a reality restructured on a phenomenological level by digital devices. The digital medium is the new ontophanic matrix of the social bond in all its forms. Thanks to the digital medium, a new modality now enriches otherphany.

Online friendships provide one of the best illustrations. Today, friendship can no longer be conceived of as experiences that take place offline only. What the Facebook experiment has introduced is not only an attempt to connect people to each other for social and marketing purposes. Beyond the unprecedented marketing uses to which most analysis takes pleasure in limiting itself (confidentiality, data resale) and which are not our purpose here, there is some real philosophical value in the concept of friendship that Facebook purveys. Thanks to Facebook, the notion of friendship extends to a whole variety of relationships that would not have deserved this label before. In the minds of Facebook's designers, the use of the term *friend* may simply have been a tactic to make adopting their service easier. Nevertheless, in fact, the term has introduced into minds and practices a new relational dimension. Should we see in it an impoverishment of friendship or, to the contrary, an enrichment of friendship?

Had we not for ourselves widely experimented for several years with all kinds of friendly and less friendly relations on Facebook, and more broadly on the internet, we would not be able to talk about this. But for us as for millions of other individuals, the record is clear and obvious: all these people—with whom on Facebook we exchange ideas, we joke, we debate, we discover common points, and with whom we end up going out for a beer—all these people really are friends, each to a different degree (this concept of degree in friendships has never been more relevant)—friends whose otherphany is different from the traditional otherphany of friendship. But they are friends all the same, in a sense expanded, extended, recast. For us, Facebook should be considered an invitation to rethink the theory of friendship through practicing it. Friendship has never been as broad and rich, alive and powerful as in the age of digital social media.[2] Aristotle used to say that it is the foundation of the "bond of the state."[3] This has probably never been more true.

Otherphany and Otherness

The digital does not threaten human essence. First, human essence is inherently technical: the anthropotechnical dualism does not hold and has never held—the belief that humans and technology are heterogeneous and incommensurate. As Nathan Jurgenson and P. J. Rey say, "Individuals and social groups have always been cyborgs because we have always existed with technology."[4] Second, the digital is only a new ontophanic stage: it makes the transcendental technicity of otherness reappear, the phenomenological factory of the other, which is too often naturalized and forgotten.

In other words, the presence of the other via digital mediation introduces an extension of the domain of otherphany. On this point, there is no contradiction between informational beings and otherness, maybe even the opposite. May we risk suggesting that? For it is the networked, informational being that gives us, not to say *offers* us in the noble sense, a new phenomenological form of presence, as the telephone did in the past. In fact, today there are a tremendous number of additional opportunities to meet an other; and this is the case whatever the features of this meeting, which always depends in the end on the interested parties rather than on the medium that makes the encounter possible. It is up to each of us to see if we want to experience it and to consider that this digital presence of the other is really a presence of the other. For me—and I have repeatedly experienced it personally in the most diverse circumstances—there is no doubt about it: it is indeed the other that is offered to me to meet through contemporary digital artifacts. Of course, those who consider that no experience of the other is possible through this medium will for sure not have an experience of the other this way. Let us be clear: if the digital object offers the (technological) possibility of an occurrence of the other, it is incapable on its own of creating the encounter. By voluntarily connecting with the digital object, the psychic subject chooses to make of this occurrence a real encounter with the other. This is why a meeting on the internet, whether amorous, friendly, or professional, "works" only for subjects who *have a priori psychologically accepted* that it could work "for real"—and not for those who "try it out" without believing (and even less for those who do not even want to try). Only those who a priori believe the possible truth of an other in the digital otherphany access the possibility of meeting using digital means. Therefore, the phenomeno-technological mediation of otherness is perhaps not just a question of otherphany, which is the phenomenalization of otherness. It may well be that otherness is inseparable from otherphany, that is, from its conditions of appearance.

Supplement 2 Ontophanic Feeling

Revised excerpt from a chapter published in 2016 in the edited volume *Vivre par(mi) les écrans*

As with a techno-transcendental structure, the technological system of a given era shapes the phenomenal aspect of the world we experience. What we must be careful to grasp and stands repeating is that it is not so much the *object* of perception that is different, but the *act* of perceiving that has changed, since by experiencing the world whose ontophanic aspect changes, it is the very way in which people *feel* in the world that has been reworked. We mean to stress the qualitatively felt experience of being in the world. From now on, this is what we will call *ontophanic feeling*, a felt sense of and a feeling of experiencing the presence of the world. The ontophanic feeling must then be considered the result of a transindividual process, which consists of a subjective, psychic dimension, as well as an objective, historical-social dimension.

The subjective psychic dimension can be likened to Dennett's theory of *qualia*.[1] What are qualia? They are the different qualitative and subjective aspects of our mental states that are ineffable, intrinsic, private, and directly accessible. Those are the "qualitative and phenomenal characteristics of sensory experiences, by virtue of which they are either similar or different from each other as they are."[2] In short, that's what we sense and feel *in a unique way* every time we perceive something. This is the first dimension of ontophanic feeling.

"Voir et percevoir à l'ère numérique: Théorie de l'ontophanie," in *Vivre par(mi) les écrans*, ed. Mauro Carbone, Anna Caterina Dalmasso, and Jacopo Bodini (Dijon, France: Presses du Réel, 2016), 63–85.

The second is the historical-social dimension, inasmuch as it results from the objective living conditions of the group at a given time. We are particularly interested in this second dimension. Our hypothesis is that within this dimension, the technological culture in which we live at a given time has a phenomenological influence on the qualia we perceive. Why? Because a being or the real is always particular and accidental, sensitive to the technological conditions of the times. Being-in-the-world and being-there (*Dasein*) are not general metaphysical conditions detached from the conditions of the century and anchored in the mind as if the mind were immutable and without an environment. Being-in-the-world or being-there are not the same thing, depending on whether you live in a dawning technological system or in a digital technological system. As we have seen, each technological system creates different ontophanic conditions—that is, material conditions of phenomenological manifestation that are its own and particular to it, as if, according to the technical reference system in which we live were not qualitatively the same the same world-as-world we experience.

This is why, following Bertrand Gille, *technological systems* can no longer be considered only higher levels of technological coherence and combination but should be thought of also as *ontophanic milieux*. We are not exactly pointing to the "technological milieux" that use the term *milieu* the way French Simondonian thinking does, but rather as perceptual environments (*Umwelt*) in line with Jakob von Uexküll's foundational work. Indeed, Uexküll's notion of *Umwelt* is not at all *environmental*, as it is sometimes mistakenly translated. From the beginning of *Streifzüge durch die Umwelten von Tieren und Menschen* (Journey into the *Umwelt* of animals and humans), after having described and analyzed the "world of ticks," Uexküll clearly differentiates between the "milieu" or one's "own (immediate) world" (*Umwelt*) and "environment" or "surroundings" (*Umgebung*). The first is the perceptual world as experienced by an animal, inasmuch as it is particular to its species and as it depends directly on the sensory equipment with which that species is endowed. The second is the "surrounding we see around it,"[3] that is, "our own human milieu" insofar as it in turn constitutes our (own) perceptual world.

And so, in combining Gille's and Uexküll's ideas, here is what we are maintaining: that just as animals live during their lifetimes in a perceptual environment that is theirs and that directly stems from their *spec*ific sensory apparatus (as in that of their *species*), so, too, do human beings live in their own perceptual environment, which qualitatively stems from the systemic technological apparatus of their time. This is what we are calling *ontophanic milieux*, which in the end is to say *Umwelt* technologies or, if you will, technoperceptual milieux. No doubt that "from Uexküll's *Umwelt* to technological medium there is a connection,"[4] but one can see here that the concept of milieu as we understand it is taken in a phenomenological and original sense, and as such it has phenomenological power that the concepts of "milieu" in Simondon's work and "technological milieu" in Bernard Stiegler do not have.

Besides, just as Uexküll invited us to imagine that each animal has surrounding it "a sort of soap bubble that represents its milieu and fills it with all the features accessible to the subject,"[5] so must we imagine that human beings in a given historico-technological era live in a kind of phenomenological soap bubble or technoperceptual bath that is profoundly unique and characteristic of the time.

To a given ontophanic milieu, there corresponds a particular ontophanic feeling made of original and singular *qualia that cannot be reproduced in another ontophanic environment.* Just as we will never know in perceptual terms "what it is like to be a bat,"[6] we (who live at the beginning of the twenty-first century) will never know what it means to perceive the world in an ontophanic context of the earliest type such as that of the Renaissance, for if for the animal "a new world takes shape in each bubble," said Uexküll, a new world also takes shape for humans in each ontophanic milieu.

Thus, each generation learns to perceive the world in its phenomenological bubble by negotiating its perceptual relationship with reality using the technical or technological devices it has at its disposal. As such, we are all *tech natives*, that is, (phenomenologically) born of a (dominant and systemic) technology that structures our psychic-cognitive connection to the world: born of the railroad and photography, born of electricity and the telephone, born of the computer and the internet, and so on.

If the term *digital native* has any meaning, it is a phenomenological meaning. For the fact of being born and raised in the digital soap bubble is phenomenologically different from being born and raised in the mechanical soap bubble. Being a digital native is therefore not a question of age but of outfitting. To be a digital native is to have acquired the faculty to watch the world appear by being digitally connected. Existence is a technoperceptual birthing before the presence of things.

To learn to feel this technoperceptual aspect of presence means accessing *ontophanic feeling*.

Supplement 3 Against Digital Dualism: A Phenomenological Critique of Judgment

Revised excerpt from a chapter published in 2016 in the collective work *Frontières numériques et artéfacts*

"Digital Dualism"

There is an ordinary metaphysics of the digital that operates at the heart of contemporary imaginary technologies and deeply nourishes beliefs that organize it. *Metaphysics* here means a theory of being that suggests a great sharing of the world. Based on an ontological division, this common metaphysics postulates that the contemporary world is divided in two by an invisible border. On both sides of this border, two substances or spheres are put in place: the first is the digital sphere/online/on-screen; the second is the sphere called physical/offline/off-screen. In this approach, *virtual* is the term we use to designate the first sphere, while *real* is what we use for the second.

For us, this metaphysics of the real and the virtual is a *profane* metaphysics in the sense that it lacks any heuristic character. It has no scholarly characteristic and produces no knowledge. Better, we must consider it an *epistemological obstacle* because it is just the more or less intellectualized magical by-product of a belief—a belief widely shared not only by most users in the digital age but also by the media, public authorities, and, unfortunately, a significant number of researchers in the humanities and social sciences who make liberal use of the distinction between real and virtual. But, dare we ask, what is the scientific basis for this distinction?

"La Fin des frontières entre réel et virtuel: Vers le monisme numérique," in *Frontières numériques et artéfacts*, ed. Hakim Hachour, Naserddine Bouhaï, and Imad Saleh (Paris: L'Harmattan, 2016), 135–146.

Let us first look at the history of ideas. In the Western conceptual tradition, as we know, there is nothing stopping us from using the terms *real* and *virtual* and then exploiting this false opposition in order to produce dualistic metaphysics that divides the world in two. Already in Aristotle[1] what is virtual or "potential" (*dunamis*) is not in opposition to what is real or endowed with existence, but constitutes another regime of reality (or mode of existence) that is defined by the fact that it stands apart from the present, that is, what is "in action" (*energeia*). Thus, "the non-present as introduced by Aristotle is by no means the opposite of reality, even though it is action that constitutes its perfection and fulfillment in all matters."[2] Two thousand years later, in the 1990s, while pioneering the question of simulation and synthesis of images, Philippe Quéau reminds us:

> *Virtus* is not an illusion or a fantasy or even a mere eventuality thrown off to the limbo of the possible. It is very real and in action. *Virtus* fundamentally acts. ... The virtual is therefore neither unreal nor potential: the virtual belongs in the order of reality.[3]

Still others, such as Pierre Lévy, expose "the easy and misleading opposition between real and virtual"[4] from the beginnings of the World Wide Web and show that on the Web, the "virtualization" of individuals, groups, actions, and information is not an unrealization but rather a "reverse actualization," by which a "deterritorialization" (from Deleuze and Guattari) or a "detachment from the here-and-now," an "outside-of" is actually accomplished. That is why with the internet what is virtual comes to consciousness as what is "not there" and is similar to the "presence of what/who is absent."[5] But "the virtual is not for all of that imaginary. It produces effects."[6] In a word, it's just as real as a presence. No way to find a reason to oppose the virtual to the real.

Now let us look at recent technological developments. Since the mid-2000s, with the combined boom of social media and mobile devices, digital technologies have kept redrawing their territory by returning users to their "being-there" (*Dasein*) in the most *local* sense of the term. Indeed, far from detaching us from the here-and-now and transporting us into an alleged, virtual behind-the-scenes, smartphones instead attached us to the most physical space (geolocation) and the most precise present (real time), to the point where the so-called distinction between the two orders (virtual and real) crumbles. As Nathan Jurgenson points out, in the age of Facebook and mobile devices, the online and the offline worlds not only are not separate but also cannot be: "Dialectically related, one can be used to bolster the other."[7] This is what the experience of the Arab Spring of 2011 shows, in which social networks played a leading role: by analogy with the butterfly effect, John Maeda coins the term *Twitterfly effect* to describe the fiery way that 140-character tweets spread via Twitter from Tunisia to Libya and Egypt, provoking popular uprisings and revolutions. This is far from the idea of immersion in a parallel virtual reality. To the contrary: "New technologies in question—especially the highly interrelated mobile web and social media—effectively *merge* the digital and physical into an *augmented reality*."[8] Thus:

> The physicality of atoms, the structures of the social world and offline identities "interpenetrate" the online. Simultaneously, the properties of the digital also implode into the offline, be it through the ubiquity of web-connected electronic gadgets in our world and on our bodies or through the way digitality interpenetrates the way we understand and make meaning of the world around us.[9]

Quantified-self applications, which have become widespread in our networked smartphones, provide a remarkable illustration: while we are jogging, for example, they record our heart rate, count the number of steps we take, generate performance charts, and share them with our friends on Facebook and Twitter. There is no way of finding in these new usages any opposition between two spheres, one of which would be real and the other virtual.

And a page has been turned in the analysis of the digital phenomenon. It is no longer possible to separate digital uses from nondigital uses: the same fields and objects are open for research. This is what a new generation of researchers, who agree with the pioneers of the 1990s, is demonstrating, simultaneously and from various disciplines. In Quebec, based on the study of online video games, Maude Bonenfant shows that "digital worlds are not 'virtual.'"[10] In the United States, from the standpoint of sociology, Nathan Jurgenson has coined *digital dualism*[11] to point out the belief that "views the digital and physical as separate spheres."[12] In France, from inside the field of philosophy, we ourselves have developed a phenomenological critique of the virtual, aiming to show that the *digital* and the *nondigital* are constantly co-constructing each other; in a fundamentally hybrid way, they form a fundamentally unique substance.

Where Does Digital Dualism Come From? Experiment in Phenomenological Archaeology

Digital dualism must not only be presented as a scientific impasse, that is, as an error. It must also be analyzed as an intellectual symptom, that is, as an illusion in the Freudian sense. To this end, we adopt here a characteristic approach that we designate as *phenomenological archaeology*. It is archaeological in Michel Foucault's sense—in the sense that it undertakes a theoretical reconstruction birthed in genealogy. It is phenomenological in Maurice Merleau-Ponty's sense—in the sense that it analyzes both usages and experiences from the point of view of perception as a primordial opening onto the experienced world. But it also feeds on the results of psychoanalysis of intellectual life (Gaston Bachelard, Didier Anzieu,[13] Sophie de Mijolla-Mellor[14]), insofar as Didier Anzieu studies "the infiltration of primary psychic processes (desires, anxieties, fantasies) into secondary psychic processes (judgment, reasoning, thought)."[15]

Phenomenological archaeology can be defined thus: the philosophical method that consists in analyzing, based on a series of lived experiences of use, how a fact is phenomenologically received into the social imaginary, that is, the way in which individuals, as a function of the cultural referents they have in common, *interpret* their perceptions to derive socially shared *value judgments*. These value judgments are not logical judgments (knowledge) or aesthetic judgments (taste), but what we will call phenomenological judgments (perception).

Digital dualism is just one good example. By "phenomenological judgment," one must understand an intellectualization of a perceptual experience, where intellectualization is a "process by which the subject tries to give his conflicts and emotions a discursive formulation so as to control them."[16] Phenomenological judgment is thus pseudoknowledge, whose main function is to try to go beyond a primary psychic experience. What then is the primary psychic experience that presides over digital dualism?

This is the trauma that digital interfaces introduced several decades ago into our perceptual habits. To perceive in the digital age is not just to perceive new objects. In the first place, to perceive in the digital age is to be forced to renegotiate the act of perception itself. It is a way of relearning the meaning of reality.

That is why we claim that the digital revolution is an *ontophanic* revolution—that is, shaking up the way beings (*ontos*) appear (*phaino*). Virtual images (in the sense of simulations), web pages, avatars, video games, persistent worlds, and social networks interaction are all new "beings" that take part in the phenomenological shock of our existence in the digital age. This shock is a primary psychic experience that can take the form of anxiety (which leads to technophobic intellectualization) or desire (which leads to technophilic intellectualization). In both cases, there is a phenomenological rupture.

So, it is precisely in order to absorb this shock to our perceptions that we *interpreted* them magically and that we began to believe in the existence of two separate worlds. We came to believe that we have a "second self" and a "second life."[17] We started doing metaphysics without knowing that's what we were doing, and believing in the division of the world in two, as in Plato's allegory of the Cave: a *real and authentic* world (the real) based on physical processes and concrete objects involving our body and a *false and illusory* world (the virtual) based on computer simulation processes and abstract software beings that cut us off from our bodies. The movies echoed this well in 1999 with the film *Matrix*, in which the virtual comes to embody in this gigantic neuro-interactive simulation, the matrix, while the real embodies the world outside it.

So there is the metaphysical dualism of the true and the false hiding behind the digital dualism of the real and the virtual. And the metaphysical dualism tends in turn to induce the moral dualism of the good and the bad.

Today we understand that we were mistaken because we are beginning to take in the new ontophanic digital culture in which we are living. For nearly thirty years, we have progressively become accustomed to the perceptual culture induced by digital interfaces, and our ability to perceive has become accustomed to virtuality. We have learned to live with computer-simulated realities and to consider them things among things: "We have learned to take things at interface value,"[18] that is, to see things as the interfaces offer them to us to see. We have entered what I call the negative of digital dualism, the "digital monism,"[19] according to which the real forms one-and-the-same continuous, fundamentally hybrid substance that is both digital and nondigital, online and offline, and on- and off-screen. Our kids experience it every day moving from tablet to paper and vice versa. All the same, they do not experience their lives as split into two separate worlds. They will soon be making fun of this old, dualistic idea maintained by their elders.

In this respect, without bearing the value of a general law, the recent testimony of Paul Miller—the young American who was trying to find his true self by disconnecting from the internet for a year—is quite edifying. In May 2013, in the New York magazine the *Verge*, he draws the following conclusions about this experiment:

> It's a been a year now since I "surfed the web." ... But without the internet, it's certainly harder to find people. ... I can tell you that a "Facebook friend" is better than nothing. ... My plan was to leave the internet and therefore find the "real" Paul and get in touch with the "real" world, but the real Paul and the real world are already inextricably linked to the internet. Not to say that my life wasn't different without the internet, just that it wasn't real life.[20]

The ultimate irony is to see this expression "real life" applied here to describe "connected life" since it is usually used to describe life offline—proof that something has definitely changed in the way we perceive what is real in the era of digital monism.

II A Short Treatise on Design

Foreword to the First Edition
Raising the Question of Design

Mads Nygaard Folkmann

Approaching the field of design requires reflection and a certain subtlety. Until recently, it was generally considered an implicit problematics of maker professionals and an integral part of practice. Over the past twenty years, however, the discipline has been the object of new approaches and is now considered a significant way for humans to interact with their environment. From the classic concept related to manufactured products (as pure products of the Industrial Revolution) created by professional designers, the concept of design has expanded to include *solutions*, with the design of systems and services or digital design, for example. The word *design* has also evolved and has now taken over in most European languages (in German, it has almost replaced *Gestaltung* or *Formgebung*). Whether as a concept or as a word, design today is almost ubiquitous.

Grounded in the latest developments in the discipline, *A Short Treatise on Design* by Stéphane Vial, a work as intelligent as it is lucid, is a major contribution to contemporary discourse. Its main objective is to ask what design is, what its constitutive elements are, and what its reason for being is. The author offers a philosophical discussion and his thoughts on design—thereby situating the book perfectly within the framework of the traditional philosophical genre of the *treatise*—thus fueling the debate and addressing issues related to design, including an ontology of design and how it can be considered a way of thinking.

Through these reflections, this book lays the cornerstone of what might be called the *philosophy of design*, a relatively new trend in design theory, even if, for twenty years (or more, if we think of Vilém Flusser's phenomenological analyses on the objects of concrete design; or Otl Aicher's thinking about the impact of design as a proposal or *Entwurf*), the latter integrates different philosophical approaches. And yet few are those who have systematically and essentially committed to thinking about design by questioning its conditions as well as its ontological and epistemological constraints, but also the ways in which it conditions the human experience. Stéphane Vial's book seeks to launch such a reflection. Some might argue that the very existence of a philosophical treatise on such a "practical" discipline as design would have been impossible until very recently. But today the field of design is ready for such an approach.

Stéphane Vial's book is an especially significant contribution to what could be called the *phenomenology of design*, an expression designating an approach to how design, in its multiple forms and its ability to create the tactile and visual surfaces of the modern world, affects, structures, and frames our experience. Regarding the role of material objects as producers of meaning, there have been other attempts of this type, including actor-network theory, which foregrounds the active role played by objects in networks established with humans, especially by orienting our behaviors. One can also find such an approach in the study of material culture, which highlights the major role the material environment plays in the development of social forms.

In this respect, Stéphane Vial's approach considers the *production of effects* that arise from the creative act that is the design process as a fundamental building block of design; as he explains in his afterword, "Design is not a field of objects, but a field of effects." In a more detailed argument, he distinguishes three types of design effects: an *ontophanic* effect to enhance lived experience; a *callimorphic* effect focused on formal beauty and the external aesthetic appeal of design in terms of form, volume, tactility, graphics, and interactive expression; and a *socioplastic* effect on shapes able to reshape society. In other words, what matters in design is not the appearance of the object but rather its ability to produce effects that condition experience. Design effects can be contained and mediated through objects that are thus potentially able to bring greater delight to our daily lives.

Relying on this way of thinking about—and deconstructing—the concept of design, this new edition of Vial's book features a new chapter on *design thinking*, which consists of thinking in, with, and through design. The author explains how designers sometimes consider this concept a marketing ploy to enable them to sell their services better by conceptualizing in the extreme, but he also insists on its qualities when it comes to designing or making innovation accessible to everyone.

With these three different design effects, Stéphane Vial highlights the reciprocal relationship among exterior shape, social impact, and lived experience, and underlines the ability of design to create, reflexively, a frame of experience. Based on the same type of argument, in his recent book *Being and the Screen* (PUF, 2013) he investigates the ways in which new digital media are changing the structures of perception, especially when they present virtual perception spaces. To approach the phenomenology of design is then to focus on understanding the relationship between design and the elements that constitute experience as well as how it transforms them.

The field of design needs a philosophical approach such as the one Stéphane Vial proposes. No longer bearing only on a specific category of objects but seen as a way of interacting with the world, design must be the object of a reflection about its very own essence: what it is, how to conceive of it, and what its implications are. Stéphane Vial asks the fundamental question of design and provides us not only with answers, but also some paths to invite us to reflect on design, and in turn raises new questions about what design is and what it implies in a contemporary context and for the future.

M.N.F.

1 The Paradox of Design: Wherein We Show That Design Thinks but Does Not Reflect upon Itself

Design keeps thinking, but it is unable to reflect on itself. It has never produced a theory of its own, as art has. Only a few instructive formulas owing to a handful of famous designers have seen the light of day. Yet, design is par excellence a thing that thinks. "For me, design is a way of discussing life, society, politics, eroticism, food, and even design," Italian designer Ettore Sottsass explains.[1] Similarly, Japanese designer Kenya Hara, Muji's artistic director, adds:

> The whole world looks different if you just rest your chin in your hand and think. There are an unlimited number of ways of thinking and perceiving. In my way of understanding, to design is to intentionally apply to ordinary objects, phenomena and communication the essence of these innumerable ways of thinking and perceiving.[2]

Trying to think about design, then, is to be caught in a liminal paradox: design is above all a thought practice, but there is yet no coherent philosophical or theoretical thought practice.[3] Not among designers; not among philosophers. The latter, bewitched by the idea of beauty since its Greek origins, have remained prisoners of a kind of "callocentrism" and have focused only on the fine arts, of which they have made a science, for which they have built a system and developed an ontology without the least concern for the decorative arts and the applied arts, considered "minor" arts. Although born more than a century ago, design is still an orphan without a theory. No "illuminating text" in "the still stammering literature about French design," no attempt to "accurately outline the field and the stakes of design," Marie-Haude Caraës rightly points out.[4] Worse still:

There is no absolute, clear, and definitive distinction between design and other fields to which it is currently connected, such as art or engineering. There is a permeability among these disciplines that makes it difficult to know where the one stops before becoming the other.[5]

The confusion is such that designers themselves end up resigned: "Design is an activity, whose lack of definition we tolerate," says Jean-Louis Frechin.[6]

Trained in the methodical division of genres and species, in Plato's method of diairesi, the philosopher that I am cannot believe it. That a discipline such as design—whose history is now clearly established; which has clearly identified professional practices and well-developed methods and tools; whose educational institutions are internationally recognized; and whose major players are known to all—can evolve now in such conceptual approximation, *that* is as stunning as it is inconceivable.

If this *Short Treatise* had only one purpose, it would be to eliminate this state of confusion once and for all and to draw a clear demarcation line between design and nondesign. That's why this book is a treatise: it attempts to *treat* design, that is, to submit it to thought. But it is also a short treatise: first of all, because I do not believe in endless works, whose length is only an opportunity to increase the author's "pleasure of thought"[7] as it reduces the reader's; second, because I do not believe that a treatise can fully treat a single question: every treatise, by definition, mistreats.

2 The Disorder of Speech: Wherein We Deconstruct and Rebuild the Word *Design*

Most people in France think of the word *design* as an adjective, and they say, "It's so design." Usually they mean good-looking, elegant, distinguished, classy; or modern, new, original, trendy, or "with it"; and sometimes they mean: quirky, bizarre, extravagant, crazy. Don't pretend not to know what I'm talking about; you know, it's like English speakers for whom it would be more *design* to say, "It's so *chic*!" The French say, "It's so *design!*" In other words, it is a factor of social distinction. It separates people into two broad categories: those who have taste, that is, those who have a "designer" apartment or house and those who do not, namely, everyone else. Thanks to design and interior decoration magazines, "TV shows that transform your interior," and websites that tell you "everything you need to decorate and lay out your house," the public at large associates design with housing, furniture, and decoration. In Carouge, in Switzerland, there is even a decorations shop called "It's So Design." As for the big names in this sector, besides displaying the name and photo of "their" designers on the shelves of their stores, they use the word *design* in their communications campaigns as marketing attractors. As I write, for example, the slogan "Beautiful Designer Kitchens" can be read at the British IKEA website. The slogan makes two assumptions: the first is that designers know how to make beautiful kitchens or that a designer's kitchen is necessarily beautiful—in other words, design has the power to create and takes on the role of creating beauty; the second is that it's eminently desirable to get a designer kitchen or that design is a value in and of itself—in other words, that design on its own signifies consumption. That means that in buying a "beautiful designer kitchen," what one acquires is not the kitchen as a product but the kitchen as a signifier—in other words, an abstract idea of designer kitchens.

To consume, Jean Baudrillard taught us, is not to make use of things but to enjoy the signs located between things and us—which triggers the act of purchasing more than the things themselves do.[1]

Thus, in the contemporary public space reduced to a space for consuming signs, design emerges as one signifier of consumption among others, staged by consumer society in consumer products, for the purpose of purchasing them. Bruno Remaury rightly indicates that design "serves to dress up a product, from a sofa to a clock-radio" and, in so doing, "to inscribe the object that benefits from it in a judgment of taste, strictly bounded by the times," that is, in the ephemerality of the "taste of the day."[2] It is synonymous with decoration and considered a beautification feature or producer of style:

> From this look *du jour* is born a decorative repertoire that is not completely "design" or even always designed, that is to say an often overburdened and unnecessary neomodernist vocabulary intended to *seem modern* and which fills shops—a zig-zaggy metallic lamp stem; trompe-l'oeil-ed chrome bolts; triangular, black plates; green and yellow toasters with big red push-buttons; strollers calibrated like off-road vehicles; stereos *à la* Goldorak. This sort of design is purely and simply a decorative repertoire meant to be bought to keep up with the times.[3]

That is what I will call *design degree zero*, or a "decorationist" conception of it to the extent that it is fundamentally rooted in introducing into consumption the idea of design, reduced to the status of exclusive predicate of the judgment of taste.

To start understanding something about design, therefore, we must turn away from decorationist incantations and first look at the origins of the word. Contrary to popular belief, the term *design* comes not so much from English as it does from Latin. Introduced into the English language from the Latin term *designare*, "to mark with a distinctive sign, to draw, to indicate," the word *design*—*dizajn*—is composed of the preposition *de* and the noun *signum*, "mark, sign, imprint." Etymologically it means "to mark with a sign," a sign having the characteristic of being distinctive, that is, the power to create difference. If the English verb *to design* etymologically "means 'de-sign,'" it does not mean "to take away its sign" as Vilém Flusser mistakenly claims[4] but, on the contrary, to mark with a sign, to draw a sign, or to sign-al. Hence, today the English verb *to design* means to draw in French (*dessiner*), that is, to draw figures, contours, or motifs—in other words, to form signs or sign forms. The act of design cannot take place outside an act of drawing. But "to design" has two meanings: that of "drawing" sketches and plans, but also that of "conceiving" according to a "plan," that is, to project according to an "intention," a "purpose," or a "concept." In this sense, to design is not just making a sign as a mark on something (a signifier), but also to forge a "project" that will be embodied in this sign, that is, to give meaning (a signified)—hence, the perpetual oscillation of the concept of design between those of "drawing" (*dessin*) and "plan" (*dessein*), which French language dictionaries love to repeat.

Nevertheless, although the term *design* has existed for centuries in English, its first known use tries to describe a new practice that dates back to 1849 and the publication in England of the first issue of the *Journal of Design and Manufactures*. The creator and director of this journal, Henry Cole, was a British civil servant, a jack-of-all-trades, the inventor of the first Christmas card, an author of children's books, a draftsman of manufactured objects, and a member of the Royal Society for the Encouragement of the Arts, Manufactures and Commerce ("The Royal Society of Arts"). In copublishing with Richard Redgrave the *Journal of Design and Manufactures*, he aims to "establish the principles of industrial production harmoniously combining 'function', 'decoration,' and 'intelligence'" and to "wed great art and mechanical skill."[5] Design's ambition is born: converging the arts and industry and bringing out "industrial designers" capable of improving industrial art by purifying forms. Nonetheless, despite the success of the Great Exhibition of the Works of Industry of All Nations, better known as the First World's Fair, which Cole organized with the support of the Prince Albert of Saxony, from May 1 to 11, 1851, at Hyde Park in London, and which celebrates British industrial power by bringing together the arts, sciences, and industry (with, in particular, Joseph Paxton's famous Crystal Palace), it would take many years for a new discipline, design, and especially a new profession, that of the designer, to appear.

As things stand, artists are not yet ready to identify, and none of them yet identifies, as a designer. Faced with the massive industrialization that England, pioneering the industrial revolution, has been engaged in since the end of the eighteenth century, artists as well as intellectuals are concerned above all with its social consequences: rural exodus, workers' poverty, work dehumanization, unhealthiness of cities, and housing. Karl Marx, a young German just thirty years of age and close to French workers' circles, just publishes the *Manifesto of the Communist Party* (1848), in which he states:

"The whole history of mankind ... is the history of class struggles."[6] At the same time, John Ruskin, he too a young Englishman just thirty years of age, publishes *The Seven Lamps of Architecture* (1849), a work in which he denounces the degradation of the worker by the machine, the disqualification of the craftsman's work, as well as the ugliness and poor quality of manufactured products. When the Crystal Palace emerges from the ground for the Great Exhibition of 1851, Ruskin sees in it a gigantic "cucumber frame."[7]

This condemnation of poor industrial taste against the backdrop of nascent socialism leads a young English artist in the decorative arts to initiate a deep renewal. A former student of Ruskin at Oxford, a convinced socialist and diligent reader of Marx, William Morris sees in the renewal and defense of the decorative arts the only means of saving humanity from industrialization, by rehabilitating the creative work of the artist using an ornamental craft of quality and, at the same time, improving the lifestyle modern society offers. In 1861 with a few others, he establishes a furniture and decoration company; he produces many textiles, fabrics, tapestries, and prints with flower and foliage motifs, which forecast a new style, art nouveau.

William Morris is not and does not call himself a designer, but he has a designer's vision. He sees in the decorative arts a way to advance modern society and then to save it from the scourge of industry. As such, one can say that he's the David Hume of design, the one who woke up production from its "dogmatic slumber." "Art brings me to my last claim," exclaims Morris in 1884, "which is that the material surroundings of my life should be pleasant, generous, and beautiful; that I know is a large claim, but this I will say about it, that if it cannot be satisfied, if every civilized community cannot provide such surroundings for all its members, I do not want the world to go on; it is a mere misery that man has ever existed."[8]

Previously considered "minor arts," the decorative arts are then called on to play a major role in the destiny of modernity. They are the ones who will give birth, after William Morris, to this "1900 style" that we call art nouveau,[9] and which corresponds to the project of creating a "total work of arts and crafts, in which everything from architecture to ashtrays was subject to a florid kind of decoration, in which the designer struggled to impress his subjectivity on all sorts of objects using an idiom of vitalist line—as if to inhabit the thing in this crafted way was to resist the advance of industrial reification somehow."[10]

And if this encounter between the decorative arts and industry is not quite the birth of design, for being *rejected* at first, it does mark its beginnings. From then on, design has a *project*: that of creating a better world. And this utopian ambition, as Alexandra Midal has clearly shown, will characterize its entire history.[11]

The term *design*, however, is not yet in fashion. Although Cole proposed using it as early as 1849, the use of *decorative arts* or *applied arts* was still preferred at the beginning of the twentieth century. Contrary to common sense, these two expressions are generally considered synonyms of *design* and, as early as 1889, even by William Morris himself:

> "Applied Art" is the title which the Society has chosen for that portion of the arts which I have to speak to you about. What are we to understand by that title? I should answer that what the Society means by applied art is the ornamental quality which men choose to add to articles of utility.[12]

What a beautiful definition of the decorative arts! Indeed, no offense to William Morris, the expression *applied arts* has a slightly different meaning, strictly speaking. Étienne Souriau teaches us that it is just another way of saying "arts applied to industry," which flourished in France in 1863 with the founding of the Union centrale des arts appliqués (Central Union of Applied Arts), which, however, a few years later becomes—hold on tight—the Union centrale des artistes décorateurs (Central Union of Decorative Artists).[13] Such terrific confusion! It's easier to understand why design in France today is taught in institutions with names as varied as écoles supérieures d'arts appliqués (ESAA), École nationale supérieure des arts décoratifs (EnsAD), or École nationale supérieure de création industrielle (Ensci). The French love making things complicated.

Nevertheless, according to Cole's wishes, design turns out to be none other than the arts applied to industry. That is why, even if it continues looking like the decorative arts, design is truly born in 1907. That year, with art nouveau reaching full maturity, a group of artists, architects, and craftsmen emerges in Munich, Germany, under the name of the Deutscher Werkbund (Union of German Work). Its leader, the architect Hermann Muthesius, is just back from a long trip to England, where he was posted to the embassy from 1896 to 1903 to report back about decorative arts and architecture in the United Kingdom. A supporter of industrialization, Muthesius sings the praises of machines and claims an alliance between the decorative arts and the standards of industry: he sees in the Crystal Palace an example of twentieth-century architecture. The same year, the architect Peter Behrens, a Werkbund member, becomes the artistic director of the AEG (Allgemeine Elektricitäts Gesellschaft), whose new products he designs with its new brands, logos, and stationery, as much as he does its new factories and housing for working families.[14] This is the first great collaboration between art and industry. Design is born. Cole invented the word; Behrens invents the thing. By proposing an extensive collaboration at the AEG among designers, artists, and manufacturers, Behrens becomes not only the first industrial designer in history, but also the one to put an end to quarrels within the Werkbund, which has become divided: the defenders of serial production led by Muthesius, on one hand, and, on the other, supporters of a return to manual and artisanal work, in the name of the freedom of the artist, led by Henry Van de Velde, the great art nouveau master.

This reconciliation of the ideals of the decorative arts, emphasizing the work of the craftsman and the ambitions of industry focused on mass production, will then flourish in the happy adventure of the Bauhaus, taking flight with Walter Gropius, the architect, member of the Werkbund, and in sync with Van de Velde's ideas. Established in 1919 in Weimar by merging the School of Fine Arts and the School of Crafts, the Bauhaus would, according to Gropius, be an educational institution and an artistic advisor to industry, crafts, and craftsmanship. Considered today—not without some idealization—as a real modern design laboratory, the Bauhaus was paradoxically centered on craftsmanship and painting. The main studios of the school are bookbinding, textiles, printing, metal, woodcarving, stone, carpentry, ceramics, and, of course, painting. The architecture studio would open its doors only much later, in the school's last years. The training was therefore mainly in the crafts, and the teaching was, rather, more the work of artists. The studios were supervised by a project manager, a technical manager, and a master craftsman, the artistic director. Among the latter were Johannes Itten, Wassily Kandinsky, and Paul Klee, as well as László Moholy-Nagy, Ludwig Mies van der Rohe, and Gropius himself. A locus of pedagogical experimentation, the school tried to open up to industry, with a goal of mass production, but its production remained mainly artisanal. The Bauhaus remains no less emblematic—even if paradoxically—of the modern movement and the functional aesthetics centered on the alignment of form and function.

So, contrary to what many have written,[15] design was not born with industry. The adventure and the consequences of the Werkbund amply demonstrate this: design is born once industry was accepted, that is, from the moment artists, architects, and craftsmen, ceasing to reject it, decided to assume industrial production and to work no longer against it and because of it, but with it and thanks to it. By "industrial production" one must understand both mechanized production, based on the use of machines, and serial production, based on the reproduction of standards enabling mass distribution. Peter Behrens thus foretells Raymond Loewy, Norman Bel Geddes, and big US industrial design agencies that start to appear in the 1930s. In a few decades, a new profession is born and becomes known to the public at large. "In less than twenty-five years, industrial world design has grown, from a hesitant, unsure probe into what *Time* called 'a major phenomenon of U.S. business,'" says Raymond Loewy in 1952.[16]

Born in England in the mid-nineteenth century, the idea of design was thus invented in Germany at the beginning of the twentieth century and comes to full realization in the United States, before being exported again to all of Europe. In France, for example, it is thanks to Jacques Viénot, who traveled to the United States to gauge the state of industrial design in order for France to benefit from his assessment (as formerly Muthesius had traveled to England to assess the progress of the decorative arts in order for Germany to benefit from his) that we owe the notion of "industrial aesthetics," introduced into the French language to translate "industrial design." After having created in 1949 the technical and aesthetic engineering office Technès, France's first design agency, of which Roger Tallon was to become the artistic director, Viénot created in 1951 the Institut de design industriel (Industrial Design Institute), which published a well-known journal of the same name and in 1984 became what today is the Institut français du design (French Institute of Design). By "industrial aesthetics," Viénot means the search for beauty in industrially manufactured objects, which is consistent with functionalist ideology according to which the beauty of a manufactured object comes from its adaptation to its function—a very "callocentric" concept that, fortunately, left room for a broader and more just vision, with the gradual introduction of the term *design* itself, which the Italians adopted very early on with the creation, in Milan in 1956, of the Association for Industrial Design (ADI), and which was finally adopted by the French Academy in 1971, after a long semantic battle throughout Europe.[17] "Strange destiny is that of words," Étienne Souriau emphasizes, "just when 'industrial aesthetics' finally acquires currency and wins the battle for good, the term is rejected and replaced by a *Frenglish* one: 'design' (and even: 'industrial design')."[18]

This term has indeed ended up prevailing, sometimes in Latinized forms, as in the Italian *disegno industriale* or the Spanish *diseño*, but without ever achieving full acceptance in the French language, whose congenital academicism regularly produces aberrations. In 1994, for example, the first Toubon law[19] provided for replacing the term *design* with *stylique*. Without success. In 2010, the Institut national de la statistique et des études économiques (INSEE; National Institute of Statistics and Economic Studies) proposes to remove the word *design* from its lexicon and translate the word by "concept" and the profession of designer by "conceiver." Absurd. It took the field of design a century to forge its disciplinary identity and professional legitimacy. No new nomenclature will be able to change that. Design is a culture embodied in a new word. The word *design* has been adopted throughout the countries of the world.

3 Design, Crime, and Marketing: Wherein We Talk about the Very Horrific Alliance between Design and Capital

> I go to the supermarket to buy milk, and I see *Star Wars* has taken over aisle 5, the dairy section.
> —John Seabrook

From the moment they started accepting industry, architects, decorators, and artists invented design. But in the same gesture, they invented designer's syndrome: a feeling of complicity with capitalism, guilty submission to the imperatives of consumer society, resigned acceptance of the market, and a renouncing of the ideal of transforming society. This syndrome appears in the second half of the twentieth century thanks to the paradigm shift at work in the industrial world: from being a society centered on the industrialization of production, now realized, in a few decades, as a society centered on the industrialization of consumption. As Alexandra Midal points out, a change of perspective occurs: "Attention paid to consumers, their choices, their tastes, their lifestyles, their values seems the designer's permanent concern." Design is now built with "fidelity to the wishes of the consumer."[1]

In *Never Leave Well Enough Alone*, published in 1952, Raymond Loewy tells the story better than anyone of the impulses behind this new way of seeing: "The finest product will not sell unless the buyer is convinced that it is really the finest," he writes. But a "competent industrial design operator knows what constitutes the consumer's picture of a fine product. He has a knowledge of the factors that are repellent to his taste or appeal to it. It is his duty to eliminate the latter and favor the former."[2] It cannot be clearer: the task of designers, and even their duty, is from now on to make products desirable in order to sell them more easily. And for good reason: "Industrial design is just as important a factor as advertising in the successful marketing of a product, a service, or a store."³ This honest talk will bring Loewy many critics, including designers who paradoxically enjoy strong commercial success, as evidenced by these somewhat provocative words of Philippe Starck, in March 2007, at a TED conference. According to him, there are different types of design: the one that we might call cynical design—design invented in the fifties by Loewy, who said: "'What is ugly is a bad sale,' which is terrible. That means that design must be just a marketing weapon, for producers to make products sexier. Like that they sell more. It is shit, it's obsolete, it's ridiculous."⁴

Amen. Yet it is misunderstanding Loewy's project designer to whom we must pay tribute as deserved: landing in New York in the fall of 1919, the same year Gropius founded the Bauhaus, the young Loewy expects to discover in America simple, slender, and silent machines and objects. Disappointed, he discovers bulky, noisy, and complicated objects: "There seemed to be a general trend to massiveness and coarseness. There was no disciplined economy of means and materials," he says.⁵ While the Bauhaus advocates aligning form and function so that all that functions well is visually harmonious, Loewy argues that "function alone does not necessarily generate beauty": complex machines like threshers or textile looms "are satisfactory, functionally speaking, but their appearance is disorganized."⁶ From then on, "It would seem that, more than function itself, simplicity is the deciding factor in the aesthetic equation," and one measures the talent of a creator by his ability to achieve simplicity by a "reduction to essentials." Hence Loewy's profession of faith: "I claim that functionalism alone could not have achieved such a result"; one must search for "beauty through function *and* simplification."⁷ Hence the elimination of the grooves, joints, rivets, and other screws to obtain smooth shapes, the rounded and streamlined features of the Streamline Moderne style. Beyond the clichés, Raymond Loewy expresses the same modern idea as the one articulated in the minimalism of a Mies van der Rohe who advocates that "less is more," or the one that is repeated nowadays in the "simplicity" of a John Maeda advocating for how to need less but get more.⁸

Nevertheless, by entering the paradigm of consumption, twentieth-century design is constantly at risk of merging with, not to mention being submerged by, marketing. The latter can be defined, from the perspective of Bernard Stiegler, as a technique of adoption: "Psychological techniques are developed to make us adopt the new products, because *a priori* we do not want them."[9] *You dreamt about it, Sony made it.* Everyone experiences this every day. Ever since then, as Benoît Heilbrunn shows,

> What marketing expects of design is precisely to project a universe of signs on its products to induce purchasing criteria that are no longer the sole province of function. ... In this sense, design seems to take over from marketing's efforts, to embed meaning in consumer goods by projecting emotional and imaginary meaning to increase their desirability and perceived value.[10]

We are far removed from William Morris. Design reduced to marketing is no more than a process of reconfiguring signs of "borrowing by modifying and re-orchestrating the codes of the product universe," rather than trying to invent new life scenarios, which is what innovation is really about. "Wherefrom this huge, anthropological *bricolage* of signs that is the market."[11] Hence, too, the unfortunate semantic shift, to which design schools have resigned themselves, which is to have "substituted the product for the object."[12]

Hence we speak of product design, not object design. This case has been settled in schools and agencies. Not content to simply brandish Loewy-type paraphrases such as "a good design is one that sells," global design agencies in the 2000s openly acknowledge their role as merchants and reinforce the confusion between design and marketing using foolish slogans: "Design is not tactical, it's strategic," reads one website, *Applause*. This means that design is the discipline that involves asking questions such as: What interest does my customer have in my offering? How to be different and create value to gain competitive advantage? How to position the product? What communications strategy to adopt? "With design, we create, reinforce, and perpetuate the intimate bonds and emotions that tie Brands to their audiences," the manifesto of this agency proudly continues.[13] Nice example of what is called "communications design," an expression that has been used for a few years now, mostly in France, to promote the creative aspect of marketing.

This is exactly how the Communication Design Association, now the Association Design Conseil (ADC), defines it: "We want to promote the goals of Design with its real ability to make a difference, to bring added value, to increase the impact of image on immediate business."[14] Such a gorgeous definition of design! I see it more as a very effective definition of marketing, whose purpose is not to create a "design effect" on individuals but to provide brands with "added value" for them. It works a bit like IKEA: just as the Swedish brand sells the "design" signifier to consumers, as in "Beautiful Designer Kitchens" (a "decorationist" sort of design), so too do global design agencies sell to brands the "design" signifier in their beautiful *branding* statements about "strategic design" (a commercial conception of design).

It therefore becomes easier to understand that American art critic Hal Foster dared to associate *design* and *crime* in the title of a renowned essay, echoing Adolf Loos's also well-known essay, "Ornament and Crime." But what is the crime? It is precisely what constitutes for design this "strategic" alliance with the logic of the market: "Design abets a near-perfect circuit of production and consumption."[15] According to Foster, just like art nouveau and its project of total artwork constituted *Style 1900*, "the universe of total design" in which we live today is *Style 2000*: "Everything—not only architectural projects and art exhibitions, but everything from genes to jeans—seems to be regarded as so much *design*."[16] But where the art nouveau of 1900 resisted the effects of industrialization in the name of the artist's subjectivity, the total design of the 2000s "savors postindustrial technologies" and spreads its hold over individuals and society: "The rule of the designer ... ranges across very different enterprises ... and penetrates various social groups."[17]

We live in the era of "the political economy of design," in which design shapes objects as well as subjects. On the one hand, the designed object can as easily be "young British art or a presidential candidate"; on the other, the postmodern individual is himself reduced to a "designed subject," which is to say, a subject shaped by communications design. As attested to by Octave's cynicism in Frédéric Beigbeder's well-known novel, *99 francs*:

> I am an advertising executive. . ... I am the guy who sells you shit. Who make you dream of things you'll never have. ... When, after painstakingly saving, you manage to buy the car of your dreams (the one I shot in my last campaign), I will already have made it look out of date. ... I intoxicate you with new things, and the advantage with the new, is it never stays new for long. ... I'm everywhere. You'll never get away from me. Wherever you look, you'll find one of my ads centre-stage. ... I decree what is True, what is Beautiful and what is Good. ... The more I play with your subconscious, the more you obey me. ... Ohh, it feels so good when I get inside your brain. Oh yes, yes, I'm coming inside your right hemisphere. What you want is no longer yours to choose: I'll tell you what you want.[18]

What a beautiful business design is! Does this mean that it has become immoral? When we think about William Morris's magnanimous project of nurturing the transformation of people and society, all we can do is question the ethics of industrial design. Does it even have an ethics? For the Austrian designer Victor Papanek, the first to so clearly raise the question of ethics in design, it seems the answer is no. In 1971, he published *Design for the Real World*, a book whose preface begins with these words:

> There are professions more harmful than industrial design, but only a few of them. And possibly only one profession is phonier. Advertising design, in persuading people to buy things they don't need, with money they don't have, in order to impress others who don't care, is probably the phoniest field in existence today. Industrial design, by concocting the tawdry idiocies hawked by advertisers, comes a close second.[19]

To the question of whether designers are complicit in consumer society, Papanek thus answers yes. To the question of whether designers have a direct responsibility in the misdeeds of capitalism, Papanek thus answers yes. To the question of whether the designer's work must be morally engaged, Papanek answers yes again. "In all pollution, designers are implicated at least partially," he writes. "It is about time that industrial design, *as we have come to know it*, should cease to exist"; "the best and simplest thing that architects, industrial designers, planners, etc., could do for humanity would be *to stop working entirely*."[20] Designer's words. That's why Papanek does not conceive of design outside an ethical approach: "This demands high social and moral responsibility from the designer. It also demands greater understanding of the people by those who practice design,"[21] for "design must become an innovative, highly creative, cross-disciplinary tool, responsive to the true needs of men."[22]

Nice point of view, but can an industrial designer escape the logic of industry? In *The Shape of Things: A Philosophy of Design*, published posthumously as *Vom Stand des Design* in 1993, Vilém Flusser extends Papanek's ideas and argues that design is subject to industry without any ethical concern. Relying on the English-language meaning of *designing* in the sense of "hatching a scheme or proceeding deceitfully," he associates the idea of design to cunning and perfidy: "A designer is a deceitful conspirator laying his traps," he writes.[23] But this ruse is above all technology's fraud with respect to nature: a lever is an artificial arm. "That is the design that underlies all culture: to outwit nature using technology, to outdo nature with art, and to build machines from which there comes a god which is ourselves."²⁴ The problem is that the sneaky and perfidious aspects of design do not stop there. "The moral and political responsibility of the designer has taken on new meaning (indeed urgency) in the present situation," insists Flusser.²⁵

First, in our postmodern societies, there is no longer any collective entity that develops standards, as religion, politics, morality did formerly, or at least these entities are very weak. "Having thus lost its relevance, any authoritative generalization of norms tends to hinder industrial progress or cause disorganization rather than provide directives."²⁶ The only instance that is still intact is science, but science provides only technological standards, not moral norms. As a result, industry is free to follow its course, without any higher normative authority to control and frame it.

Then, because after the growing specialization of industrial production, design itself has become a complex and divided process involving multiple actors: "the design process is organized on an extremely cooperative basis. For this reason, no single person can be held responsible for a product anymore."²⁷ A widespread moral irresponsibility flows from this logic of production. It's not anybody's fault, because it's everyone's fault. Thrown into "a situation of total irresponsibility towards acts resulting from industrial production,"²⁸ we can only witness the hatching with impunity of "morally reprehensible products" and objects without legitimacy. When one thinks, besides, that the progress of ordinary design (for example, designing a knife that cuts meat well) is always due to advances in military design (for example, designing a knife that slits a throat well), then "one has to make a choice to be either a saint or a designer," Flusser writes ironically. "Whoever decides to become a designer has decided against pure good."²⁹

4 Beyond Capital: Wherein We State the Moral Law of the Designer

Design is based on a structural and historical contradiction. On the one hand, design is a socialist invention: it emerges in England from the revolt against the ravages of industrialization on people. On the other, design is a capitalist invention: it emerges in Germany as mass industrial production grows in acceptance and grows up as *industrial design* in the United States. This structural contradiction is unique in the world: no other activity crystallizes to this extent such political ambivalence in its very definition. To be at once socialist and capitalist, that is what is required of the designer. That's not just a paradoxical demand but also a contradictory one: it's about doing industrial design without creating industry. It's a bit like asking someone to prepare a meal and asking him or her not to use kitchen utensils. It's just not doable. In psychology, ever since Gregory Bateson, this is called a "paradoxical injunction."[1] Based on the "double bind" mechanism, the paradoxical injunction is characteristic of the mode of functioning of schizophrenic families. According to psychoanalyst Harold Searles, it's akin to "an effort to drive the other person crazy."[2]

This is why the attitude of expecting designers to design industrial design without compromising themselves vis-à-vis the logic of the industry can be considered a form of coercion to madness, a kind of effort to drive designers crazy. So you have to choose: to design or not to design. But if we choose design, we necessarily choose industry, that is, its production tool and the circuit of distribution and consumption associated with it. There is no other way.

Therefore, "the question is no longer about criticizing consumer society."[3] In fact, it's time to learn to situate design efforts *beyond* capital, which is certainly its means but cannot be its end. Whenever the medium and the end of design have been confused, design has dissolved into marketing. Designers themselves observed this first. As early as in the 1930s, for example, MOMA's art critics denounced the work of the Streamline Moderne's designers, of which Raymond Loewy is one of the major exponents, because they saw in this new generation of industrial designers "collaborators of capitalism" who were drawing pencil sharpeners just as they were aerodynamic planes.[4] They opposed Good Design to it, which, as defined by Edgar Kaufmann in 1953, was based on a functionalist aesthetics and required being truthful with regard to materials and manufacturing just as much as to price and use.

Designers are no fools. They can tell that their discipline risks falling into the pitfalls of marketing at any time. They can see this so well that some of them, in Italy in the 1960s, even use it as the grounds of protest to found a new approach, that of critical design. Joe Colombo's Anti-Design, presented in a manifesto of the same name and published in *Casabella* in 1969, is the best example of this: he advocates abandoning the production of objects in order to stop blindly serving capitalism. According to him, design can do without materiality and must become attached, by renouncing the worship of the object, to reinvesting in space and home, completely autonomously from architecture. Similarly, Andrea Branzi, figurehead of Italian Radical Architecture, sees in design a way of reconfiguring the political space of the city, still with the idea of fighting the hegemony of the materiality of the object in an era of mass consumption. Victor Papanek, as we saw in the previous chapter, even goes so far as to denounce the perfidy and futility of industrial design, and invites designers to think about the "real needs of human beings" such as ecological and social problems.

But beyond critical postures, design cannot get rid of the market without straining to go crazy. The first one to realize this was Italian designer Ettore Sottsass, famous for his collaboration with the Olivetti brand. In 1973, two years after Papanek's *Design for a Real World*, he rebelled very ably against the systematic suspicion designers were subjected to. In a famous text titled "Mi dicono che sono cattivo" (Everyone Says I'm Bad[5]), he writes:

> Now everyone says I'm very bad; they all say I'm really bad because I'm a designer; they all say that I should not be doing this work—that I am bad—they all say that doing this work is at best to be dreaming. They all say that designers have "as the sole and only goal that of joining the cycle of production-consumption." They all say that designers do not think about class struggle, that they do not serve the

cause of the people, and that, on the contrary, they work for the system ... and that the system consumes them, digests them, and only feels better for it.[6]

This would be design's "original sin," he continues, in whose name we can accuse designers of every ill:

> I suppose I am responsible for the number of people killed on the roads since it is Capital that manufactures the cars. And it must also be my fault that city dwellers commit suicide, that love stories end badly or do not take off, that there are sick children, famines, diseases and, more generally, that all these misfortunes exist. It really seems that I must be responsible for everything since I work for industry.[7]

Such irony has the merit of denouncing a situation that makes of the designer, as rightly pointed out by Alexandra Midal, "the scapegoat of critics of capitalism," at a time when "the criticism of design dissolves into that of consumer society, marketing, advertising, and mass media."[8] An easy scapegoat since it is obviously not in a designer's power to reverse the class struggle, to destroy Capital, or simply transform the mechanics of global economics.

Let's not, however, pull the wool over our eyes, with all due respect to Mr. Sottsass. If a designer is not wicked by necessity, he is by contingency. In other words, it can happen to him, and everything depends on him because being bad or not is a choice that remains available to him and that he chooses or not. We saw precisely this in the previous chapter: marketing design is a reality, an observable reality, a stated reality, an accepted reality. When we say "marketing design," we mean that a design activity considers the market both a medium and an end. Now if there's an ethics of design, it can only rest on the following principle: limiting the market to the role of simple expedient, instrument, process—that is, one means among others— without ever making of it the goal, the objective, the *design*—that is, an end. From then on, one can, without exaggerating, erect into the moral law of the designer the following imperative, formulated in a Kantian way: *Act in such a way that you treat the market, as much as a designer in the design projects you offer users, always simply as a means and never at the same time as an end.* Otherwise, design is blind, and designers oscillate between general irresponsibility and paradoxical madness.

Besides, there is no design without morality, in the noble meaning of the word. The problem of the morality of design is an integral part of the question of design's identity. That's why "designers have never ceased, since the origins of the discipline, to legitimize its action, as if in repentance for an original sin."[9] To be a designer is to take a moral position. It is, for example, to take a position on the question of the market, as in the past at the time of the Werkbund it was to take a stand on the question of standards. This is why design is constantly looking for legitimacy: the market alone gives it none. The market alone gives it only the *means*. So designers need to find its *end* elsewhere than in the market; otherwise it has no reason for being. And at the same time they have to use the power of the market, *without which it can't exist*.[10] That is why, from a historical point of view, as we have seen, design does not emerge with industry but with the *acceptance of industry*. It is only starting with morally *assuming* the tools of production (mass production) and morally accepting their inseparable symmetrical opposite that is consumer society (mass distribution) that designers enter into the era of industrial design. Sottsass is therefore correct in adding to his pamphlet the following remark: "The problem [is] not whether we are bad or not because we are designers, but rather whether we know what we are able to do with that as designers."[11] Which brings us back to the problem of morality, the problem of choice—and what to do with it. Except that today the choice can no longer be, as at the time of the Werkbund, that of being "for or against industry" or "for or against the market." Believing yourself obligated to such a choice, as a designer, is to strive to drive yourself crazy. Just as the Werkbund ended up accepting standards, so too did contemporary design end up accepting consumerism. In order to see better beyond it and to think beyond capital.

5 The Design Effect: Wherein We Reduce the Quiddity of Design to Three Criteria

Someone has drawn everything around us. Someone drew the places where we love, work, and die. Someone drew the objects we value, keep, and give up. Every drawing concerns every aspect of life, from our living room furniture to our business offices; from urban spaces to hospital rooms; from public transportation to classrooms; from cooking utensils to heaters. It is not, however, because an artifact was the subject of a drawing that it was also the object of design work. Craftsmen draw, engineers draw, technicians draw. All the same, they are not designing. We exaggerate when we say that "everything depends on design,"[1] as Vilém Flusser claims, or that we live in a "universe of total design," as Hal Foster has it. Although everything is potentially about design, everything is not design—and for good reason: whereas design cannot do without industry, industry can do very well without design. Numbers of products and services are made without any design process. Design is not inevitably integrated into industrial production. In France, for example, it is even less the case than elsewhere, if we are to believe the most famous of our designers: "French industries are not interested in design, it's not part of their culture, which is completely different than Italy, which integrates it as a natural thing,"[2] says Philippe Starck. While design was born as soon as it was accepted in industry, industry doesn't necessarily accept design, which is proof that design is not to be confused with industrial production. It is more a kind of supplement to the industry that appears only under certain specific conditions. The problem

---

Some of the ideas in this chapter have since been published in "The Effect of Design: A Phenomenological Contribution to the Quiddity of Design Presented in Geometrical Order," trans. Nick Cowling and Marie-Noëlle Dumaz, *Artifact* 3, no. 4 (2015): 4.1–4.6.

is knowing which ones. What is it that confers on a space, a product, or a service the quality of design? How does one distinguish a designed object from an ordinary industrial object, a crafted object, or an art object? In a nutshell, what is the essence of design?

This is a question that only a philosophy of design can answer. Just as the philosophy of art ponders, at the ontological level, under what conditions an object becomes a work of art, the philosophy of design must question, at the phenomenological level, under what conditions an object becomes a *designed object*. One has to distinguish between "when one talks about design as a practice ('the practice of design')" from when one speaks about it "in reference to an object ('a designed object')," and from when one speaks of it "as a qualifier of taste (say, 'a cool design')."[3] What we're talking about here is design in reference to an object, which is to say design as a constitutive essence of a category of beings. Asking what the essence or quiddity of design is amounts to questioning the criteria at the border between design and nondesign.

I propose calling what design labors to do *design effect*. By that I mean that design, before being a space, a product, or a service, is primarily an effect that occurs in a space, a product, or a service. This means that design is not a being but an event, not a thing but an echo, not a property but an impact. As Kenya Hara aptly says, design does not consist in conceiving of "things that are" but rather of "things that happen."[4] Design indeed has a performative quality about it: before being a thing, it's *something that happens*. In Kierkegaardian terms, one could say that it is "reduplication":[5] it speaks by making itself or, better yet, by becoming. The being of design is that of becoming. To be, for design, is to happen. That is its phenomenality. Contrary to what one might believe reading too fast, the *effect* is not to be understood as a logical concept meaning "consequence" (correlated to a cause); *effect* here should be understood as a phenomenological concept to mean a creative hatching of appearance, an inventive dynamics of manifestation (correlated to perception), as it structures experience. In the scope of my existence, then, design is responsible for what presents as an experience-to-be-lived[6] (insofar as it has been intentionally constructed). This first dimension of design effect is what I first called the *experience effect* in order to emphasize the fact that one experiences design, lives it, tries it. That's in fact how one recognizes it. Where there's design, users immediately feel its effect, because their experiences are instantly transformed, improved. Take the example of the iMac, an infinitely simple idea, a stroke of genius. The iMac is a desktop Apple imagined with only a flat screen—no tower, no external CPU. Everything is in the screen, the motherboard as well as the hard drive and the read-write CD/DVD. Just plug it in and put it on your desk, and you'll instantly save some space at home. What was initially a simple personal computer triggers a new way of furnishing your space and produces effects even in your home.

The value of the experience is therefore at the heart of the design. Where there is nothing to experience, there is no design effect. Design does nothing more than generate *experiences-to-be-lived*, whether through consumer goods, urban installations, or digital services. What it alters is the *quality of the use experience*, for I can make use of a bathroom, a watch, or a telephone without it offering any quality of experience. In this case, I experience raw use: water flows in a shower, a needle displays seconds on a clock, a strident ring announces a phone call. But if I can experience sensuality using my bathroom, if I can be delighted using my watch, or if I can have fun using my phone, then I will experience movements of pleasure among the most commonplace of my actions and add quality to my life experience—hence, the priceless added-value that design can bring to industrial production and, more generally, to all devices involving users. Such is the meaning one needs to confer on the idea of design charming one's life: to increase the quality of lived experience, whatever it is. That's why I will now call this first dimension of the effect of design *ontophanic effect*.[7] Design does nothing more than change the qualitative regime of one's experience of being, that is, to be present-to-the-world, playing on how being (*ontos*) appears to us (*phaino*). It intentionally offers new ontophanies that are the subject of new experiences-to-be-lived. That's why design is not about a field of objects but indeed about a field of effects.[8]

Nevertheless, though essential and foundational, this first ontophanic dimension is not enough to determine what actually characterizes a particular design effect, since one finds it readily in other forms in works of art and artifacts in general. For a design effect to take place, it must also produce a callimorphic effect. By that, I mean a formal *beauty effect*. To design is first of all to create shapes (whether they are spatial, volumetric, textile, graphic, or interactive)[9] and seek to give them style, character, and expression. Where there is no elegance or refinement in the line, no volumetric purity or weighting, no poetic shape or perfection in the drawing, no visual seduction or graphic allure, in a word where there is no harmony of forms, that is, callimorphism, no design can occur. Design begins with the inherent enjoyment in the perception of formal beauty. This is not trivial or incidental. Seeking out beauty reflects a fundamental psychic need of human beings. As Freud indeed showed, the perception of formal beauty provides a "premium of seduction," or an extra pleasure, that we substitute for the drive satisfactions we must give up in real life. We need it to better endure or charm our existence. Design therefore plays a vital role in postmodern societies. Design is what assumes responsibility for satisfying our basic need for beauty, since "beauty has moved over to industrial technology and emigrated from the field of art, now freed from its constraints,"[10] as Jean-Pierre Séris rightly points out. That's why, in France, we have long wanted to define design as an "industrial aesthetic," considering (sometimes excessively) that the search for beauty in industrially manufactured objects was the heart of the design.

And that is why the history of design is so inflected by battles about style, like so many battles about how to think about the beauty of forms. What is art nouveau, if not a callimorphic theory based on the notion of ornament? "Ornament completes form. It is an extension of it, and we recognize the meaning and justification of the ornament in its function. This function consists in structuring the form and not in 'adorning' it," says Henry Van de Velde.[11] What is functionalism, if not a callimorphic theory according to which beauty comes from adapting to function under the guise of rejecting decor? "'Ornament' was formerly the way to say 'beautiful.' Today, thanks to the work of my life, it is a way to say 'of lower value,'" writes Adolf Loos.[12] What is modern architecture if not a callimorphic theory in "five points"? Five points that are, as Le Corbusier has formulated them, pure concepts of form: piles, roof terraces, open plans, lengthwise windows, and open facades. And what is Streamline Moderne if not a callimorphic theory based on aerodynamics? And so on—until today. Today, the dominant callimorphic assumption is that of purity and lightness. One sees it with all designers and in all specializations: in product design, as with Philippe Starck's totally transparent Marie chair made using a single mold; in graphic design, with the aesthetic of emptiness of a Kenya Hara choosing to place the Muji logo at the middle of the horizon;[13] in spatial and architectural design, with the delicate and fluid constructions of a Kazuyo Sejima, such as New York's

New Museum or the Rolex Learning Center at Lausanne's Federal Institute of Technology.

Finally, and this is its third dimension, the *design effect* always and necessarily has a *socioplastic effect*. By that, I mean a social reform effect. Creating new material shapes is about recasting life's social forms and at the same time inventing new ways of existing, together and side by side. This is made possible by the fact that shapes that arise from design, unlike those that arise from art, have a use value, that is, some material utility. They are put on the market to meet needs, and as such they circulate in different sectors of everyday life through the whole social body. Design starts where there *already* exists a use value. The use value, however, does not in and of itself create a design effect. It is simply a precondition anchoring the possibility of a *socioplastic* effect, which has been sought since the origins of the discipline. Recall William Morris who had placed in the decorative arts the hope for a social revolution capable of saving workers from the misery of labor and the artist from machine alienation in order to produce a quality living environment for all. This desire to transform society and create a better world is the utopian heart of design. It assumes that the shapes created by designers are not just plastic forms but actually *socioplastic* forms, that is, forms capable of acting on society and reshaping it. Besides, designers have always attributed to the shapes they claimed all kinds of psychological and moral powers likely to influence the human environment. Loos, for example, writes, "Not only is ornament produced by criminals but also a crime is committed through the fact that ornament inflicts serious injury on people's health, on the national budget and hence on cultural evolution."[14] Its disappearance should accelerate the cultural evolution of peoples and bring them into adulthood!

If design is above all a matter of theory of forms, this theory of forms is therefore always at the same time a theory of humanity and society. Design is always a *sociodesign*, a creator of civilization, which seeks to work on "sculpting the social."[15] And that is where its moral foundation lies. Not taking the means for the end, as we saw, is the only way for design to register its effort in the service of humanity. If the market is its privileged means, then its most essential end is to work toward social sculpture beyond capital. Seeking to improve our living environment, to put together other ways of living, to imagine new ways of being together, to face the great problems of the future: these are some of design's real issues. Alain Findeli summarizes them very well as follows: "The end or purpose of design is to improve or at least maintain the 'habitability' of the world in all its dimensions."[16] One cannot not say it any better than that.

This is why a project like Vélib', the self-service bike-share program in Paris, offers a beautiful example of *design effect*: beyond the obvious *callimorphic effect* as a result of the bicycles' slender shapes and of their stations (the latter designed by Patrick Jouin), Vélib' produces a considerable *socioplastic effect*: it changes the way we experience the city, the way we see it and move about in it, and the way we are seen in it. If we add to that that Vélib' works by using a dedicated mobile app that lets the user know in real time and using geolocation about the availability of bicycles at different stations, then we get a service that remarkably transforms urban experience and mobility, that is, the city's habitability.

6 Drafting a Project: Wherein We Show That the Designer Is Not an Artist

Art makes questions. Design makes solutions.
—John Maeda

At a time when design is exhibited in museums, when star designers are presented as "creators," and when fine arts schools are establishing design departments, one might be tempted to think that the line between art and design is disappearing or that design is beginning to look like a contemporary form of art. From an aesthetic point of view, it is true that the reception of design objects is comparable to that of works of art, as evidenced by Gerrit Rietveld's Red and Blue Chair (1917), which he conceived as an abstraction for the mind more than as a seat for the body, or, more recently, Philippe Starck's famous juicer, Juice Salif (1987), which is so strongly valued as to be put on display on a shelf rather than be used to (badly) squeeze an orange.

Yet the process of artistic creation differs fundamentally from the process of design: creating art and creating design are two very distinct things. What does an artist do? "He creates a world of his own or rather rearranges the things of his world in a way which pleases him," writes Freud. By this, one must understand that he creates an imaginary internal world, a "phantasy world" in which "every particular phantasy is the fulfillment of a wish, a correlation of an unsatisfying reality," as in dream activity.[1] From the psychoanalytic perspective, the artist is responding to a psychic functioning of the narcissistic type. Dominated by the pleasure principle, he invests the greater part of his libido in the self and, by way of sublimation, finds the means to satisfy his impulses only in his internal, psychic activity embodied in the forms he produces. In this respect, he enjoys absolute freedom and is not accountable to anyone; he has no obligation to others. All that counts is his desire, in whose name every extravagance is permitted, as evidenced, for example, by the anatomical performances of German artist Gunther von Hagens, in whose exhibit human corpses are preserved by means of his "plastination" technique.[2] That is, art has no limits (for which it is often criticized and always forgiven) as if, in the name of art, one had every right.

Designers, on the other hand, do not enjoy freedom without borders. They are subjected to a complex beam of constraints and norms that are constantly evolving. And they are especially subject to their users' verdicts. They do not work only from their own desires (a condition that remains necessary for all creative work) but also from the desire of the other. They respond to an object type of psychic function: subject to the reality principle, they cannot be content with investing their libidos in their egos and must deal with these "objects" that are external to them, who are the other. They labor in service to people and, as such, have a responsibility to others. "An artist can choose whether to be responsible towards other human beings or not, but instead a designer *has* to be, by definition" says Paola Antonelli.[3] That's why designers must always justify their approach and explain the legitimacy of their work: "The project is a complex process of objectifying subjectivity through image and discourse, which proposes and exposes, explains, rationalizes, and legitimizes."[4] "The too obvious and embarrassing subjectivity that presides over the act of creation"[5] cannot suffice when it comes to engaging others. Where artists need provide no explanation to justify their choices of shapes, colors, or materials, designers must, on the contrary, give reasons so their choices can be objectively recognized as choices that make sense to others. Without that, they result strictly from the arbitrariness of their desire, in which the users cannot recognize themselves because they are not the target, from which stems the necessary generosity that must preside in the approach so well defined by Patrick Jouin:

> A designer is someone who is curious, curious about techniques, curious about uses, curious about the behavior of others as well as his own. It's someone who has to inject elegance into useful objects, some poetry, ensuring that every moment of life is a moment of exception. It can be by simply lifting a cup; it can be using cutlery; it can be shutting the door of a car. Our job is to make all those moments

quality moments: that the objects not burden us, but quite on the contrary that they reveal what is best in us.[6]

The difference between art and design therefore creates no ambiguity. Designers themselves, who need to clearly delineate the field of their work, manage to articulate it. So, Kenya Hara writes:

> Art is the expression of individual will ... whose origin is very much of a personal nature. So only the artist knows the origin of his own work. ... Design, on the other hand, is basically not self-expression. It is born of society. ... The essence of design lies in the process of discovering a problem shared by many people and trying to solve it. Because the root of the problem is within society, everyone can understand plans for solutions and processes for solving the problem, in addition to being able to see the problem from the designer's perspective.[7]

Where artists tackle a problem that is in one way or another always personal to express, designers confront social issues to try to solve them. Their approach is profoundly socioplastic, and their desire is a desire for the desire of another.

That's why the designer does not create "works" but rather conceives "projects." A project is a set of formal original proposals that structure uses and offers users an experience-to-be-lived in such a way as to satisfy their needs and be likely to improve the quality of their lives. It's a process of projecting and anticipating that requires imagining from the existing state of innovative forms of life and use. In this sense, the designer is the one who proposes a projected state of reality, that is, a potentially realizable future. The designer is a projector. He puts forward a plan, a design, an intention. He has a vision of the future. This ability to project is what distinguishes him from any other specialist in drawing, whether he be an artist (draftsman) or engineer (technical drawing). *Projecting* in design has a very special meaning. It consists of premeditating something, of fomenting an ideal, of working on the stuff of the future. Design occupies a consubstantial area with the future. In 1969, in *The Sciences of the Artificial*, Herbert Simon was the first to point this out: "Everyone designs who devises courses of action aimed at changing existing situations into preferred ones."[8] Alain Findeli is therefore right when he points out that

> the look that design has on the world is *projective*. Let us understand by that that for the design-researchers, the world is there to perfect, it is a project and not just an object that must be described, whose causes must be explained or meanings understood.[9]

So design is a creative practice oriented to the future and resting on an intention to improve things. It's about putting oneself at the service of improving living conditions, as well as the quality of life as experienced. This is why the notion of "project" is so conclusive in design, for the project underway is the future searching for itself. So a design *projects* in the sense that it throws forth before us an ideal that must be implemented immediately and concretely.[10]

7 Design as "A Thing That Thinks": Wherein We Defend the Concept of "Design Thinking"

In recent years, a handful of Americans who claim to "dig deeper beneath the surface" are inscribing their work in an effort to *think design*, which in many ways echoes this book. They are not philosophers but designers and are led by Tim Brown of the well-known IDEO agency. Their approach is first of all that of an agency that sells services to its customers (which we cannot ignore), but their ideas are not without theoretical reach for the entire design field and could well contribute to the principles of a philosophy of design. At any rate, it seems their principles are influencing the whole world's design schools, since more and more of them are offering courses of study in *design thinking*. But what does *design thinking* mean?

To talk about design thinking is to try to think *through* design precisely as a thought. The expression *design thinking* should be understood as short for "design thinking process," which means the process of thinking that is characteristic of design. What that suggests is that design is primarily a thinker's practice or a method of thought (thanks to which it can be brought closer to philosophical activity). In this sense, one can say that design resembles the Cartesian definition of the subject, understood as existing because of the very fact that he thinks. "I think, therefore I am" becomes "I think, therefore I am a designer." Like the self in Descartes, design exists only then because it thinks. It is therefore, par excellence, "a thing that thinks" or a thinking thing.

But what kind of thinking is this about? Concretely now, how does *a designer think*? How does one define this process of thought that is specific

---

This chapter was first added to the 2011 Swedish translation. It was later revised and integrated into the second edition.

to design? For Tim Brown, the principal craftsman of the concept of design thinking[1] since the mid-2000s, it is nothing less than a new creative methodology with three essential steps.[2]

The first is about observing the needs of people ("human needs is the place to start"). Above all it's about making life easier and more enjoyable, not about good ergonomics ("putting the button in the right place"). One ought to hear culture and context. It is the moment of *inspiration*: a designer is first and foremost someone who is inspired by users, who tries to empathize with them, who puts himself or herself in their place to try to think like them and to perceive the world from their (bodily, emotional, cognitive, social, cultural) point of view—not to better sell them one's products but to better understand their needs and essential expectations.

The second step is that of experimenting as a means of generating ideas ("learning by doing"). Unlike what we typically believe, concept in the creative process does not come before realization, but always after. To design is not just thinking to make; it's also making to think. The French philosopher Alain used to say as much in 1920 about the artist: "The idea comes to him as he creates; it would even be rigorous to say that the idea comes afterwards, as it does a viewer, and that he, too, is a spectator of the birth of his work."[3] For the designer, it means making many prototypes before even thinking of proposing an idea. This is *ideation* time: the prototype returns something we will learn from it, and that will give birth to an idea. And the more designers experiment, the more ideas they will have.

Finally, the third step is *implementation*. Tim Brown says design is too important to be left in the hands of designers. At the end of the design process, we must not have only (passive) consumption but (active) participation too. It is time to implement, and design must become a participatory system because it is when design is in the hands of as many people as possible that it has the greatest impact. The whole world must get involved in the new choices that we have to make. It is therefore incumbent on design, as well as each of us, to raise questions to which designers must bring answers.

And for good reason: we live today in times of big changes, and existing solutions are becoming obsolete—hence, the need for design to change paradigms and move from a model of consumption to a model of innovation. For designers, this implies a methodological conversion: moving from *design* (marketing, decorationist, aesthetician) to *design thinking* (innovative, participatory, social). That means being less interested in the object and more in its impact (what I called its "effect," considered here in its "socio-plastic" dimension). That means forgetting the twentieth century's "small vision of design," which engaged mainly in designing more attractive and more sellable products and whose impact on the world was of minor consequence, to embracing instead a more ambitious vision of design in the twenty-first century by going after bigger issues such as education, health, safety, water, with much greater prospective social impact. These can be called human, social, or cultural problems insofar as they are present in any design project.

Tim Brown tells us that twentieth-century designers were "a clergy of people wearing black turtlenecks and designer glasses who worked on small things" centered on aesthetics, image, and fashion. Twenty-first-century designers must become system thinkers who reinvent the world using a human-centered-design logic. This apparently demagogic phrase is still fully legitimate for a design philosophy.

For sure, one can criticize the concept of design thinking, pointing out that it is not a philosophical concept but more of a marketing concept. After all, a good creative slogan that states the obvious about the creative process as if making some miraculous discovery is not a good strategic slogan. That's what Don Norman at first maintained—he considered design thinking a "useful myth":[4] useful because it's a way "to convince people that designers do more than make things look pretty," but above all a myth because what one legitimates by using this term is what creatives of all disciplines have always done. In addition, from the point of view of French culture, considering design "a thing that thinks" is nothing very new. It's what French design schools think of as the society-centered thinking of any design project, which has been associated traditionally with philosophy and the humanities in French design curricula.

One cannot, however, neglect this: design thinking is the only true design concept that contemporary, professional designers have been able to come up with to define and theorize what they do. When design itself tries to think, it joins the philosophical gambit. Does design thinking articulate something rather than nothing? I maintain that it states a truth about contemporary design that is consistent with practices and gives it meaning that is consistent with values. In just a few years, design thinking has largely proven itself and found prestigious supporters, among them

Bill Moggridge: "The 'Design Thinking' label is not a myth," he writes in response to Don Norman. "It is a description of the application of well-tried design process to new challenges and opportunities, used by people from both design and non-design backgrounds,"[5] to the point where Don Norman himself recently revisited his early criticism: "I've changed my mind: Design Thinking is really special." Rather than a "useful myth," it is in fact an "essential tool" in the sense that it is "a systematic, practice-defining method of creative innovation."[6] One can thus consider that beyond its marketing uses, the concept of design thinking has real epistemological value and must be taken very seriously. The contempt it still widely faces in France attests, beyond the ever more obsolete persistence of a culture of aesthetics in the applied arts, to the depth of French arrogance, which those who travel easily recognize. The French would do better, though, to be inspired by it, especially since among our American friends, design thinking is now unanimously used in both design agencies and among researchers in the academic world! May these few pages help to familiarize French design culture with the method of design thinking: it has the huge advantage of making it possible for not just a few isolated and well-educated individuals but for everyone to design and innovate.

8 Toward Digital Design: Wherein We Look into the Consequences of the Interactive Revolution

> Unfortunately, we were born on the wrong side of the screen. We are not made of bits, we are made of flesh, blood, and atoms. We live most our lives in a physical world, which is hard, obtuse, not adaptable, without magic.
> —Violet.net

> I watched the process whereby my daughters gleefully got their first email accounts. It began as a tiny drop—emails sent among themselves. It grew to a slow drip as their friends joined the flow of communication. Today it is a waterfall of messages, e-cards, and hyperlinks that showers upon them daily.[1]

So wrote John Maeda, digital designer, former researcher at MIT, and director of the Rhode Island School of Design, in 2006. For all those who are not digital natives, this anecdote will certainly evoke many others, and for everyone, it illustrates well the fact that since the appearance of the first computers in the 1940s, our civilization has been engaged in a profound technological upheaval: the "digital revolution."[2] Understand that this upheaval is not just a technological revolution; it also operates as an anthropological revolution. It affects human beings as much as machines. In other words, the digital is not just a technical fact: it is a "total social fact" (in the Maussian sense) inasmuch as it is an intimate, psychic fact. It is therefore important to understand how designers and architects have reacted to this revolution. How does the digital affect the field of their practice? Does the digital transform the idea of design? What is "digital design" now?[3]

This chapter was originally presented during of the PraTIC Study Day, June 1, 2010, Paris, at Les Gobelins, L'école de l'image, on the theme Le Design numérique: Discours et réalités.

With the powerful rise since the 1980s of computer tools in professional practices, one can say that designers and architects fall into one of two camps: those who, while forced to adopt digital technologies in their profession, would prefer to maintain a certain distance from them and even mistrust them, and others who are excited about new possibilities and see in the digital new creative challenges and a way forward for design.

The first person to note this ambivalence is the American psychologist and sociologist Sherry Turkle, a researcher at MIT and a pioneer in digital research. She prefers to speak of *simulation* to describe the phenomenon of computer technologies insofar as they are virtual technologies, that is, a visual simulation leading to an organization of life "on [the] screen."[4] With the launch of Project Athena in 1983, Turkle undertook an ethnographic survey of its reception among MIT students and faculty. Sponsored by large computer firms such as IBM, Project Athena aims to integrate "modern computer and computational facilities into all phases of the educational process."[5] This was the beginning of computer-aided design (CAD), which from the outset triggered two types of contradictory reactions at MIT: an enthusiastic adoption by those, mostly students, who give in to simulation's desired immersion and who start *making* with computers (the makers), and the worried skepticism of those, mostly faculty, who express great distrust for these new tools and question their relevance, fearing a loss of reality (the doubters).

In a brilliant book, *Simulation and Its Discontents* (2009), Sherry Turkle presents the results of this survey conducted in three departments at MIT: architecture and urban planning (with the beginnings of Autocad software launched in 1982), civil engineering (with the Growltiger software), and physics and chemistry (with the Peakfinder software). Using vignettes drawn from her interviews with faculty and students, Turkle shows that when CAD arrives, the *discontent* expresses itself in the same way in these three departments. Each time, even as the new tools are adopted, there is also a wish to preserve "sacred spaces," that is, work areas that are not available for digital work:

> Architects wanted to protect drawing [by hand], which they saw as central to the artistry and ownership of design. Civil engineers wanted to keep software away from the analysis of structure; they worried it might blind engineers to crucial sources of error and uncertainty. Physicists were passionate about the distinction between experiment and demonstration. They believed that computers did have

their place in the laboratory, but only if scientists were fluent with the details of their programming. Chemists and physicists wanted to protect the teaching of theory.[6]

Twenty years later, in the 2000s, after a second survey, Turkle observes that discontent has decreased considerably as researchers are led to systematically engaging with screens and "working almost full time in simulation."[7] The distinction between partisans and skeptics has gradually disappeared, giving way to a strong feeling of ambivalence, split between the happy awareness of what has been gained and the anxious awareness of what has been lost. Among architects and designers in particular, even though computer-aided design has become commonplace in professional practice, it has also become commonplace for them to define themselves by putting forward "what they do *not* do with computers."[8] Some of them, for example, complain that the computer's logic, to which they must bend to do their work, inhibits creative thinking. Others regret seeing their work as designers reduced to simple options in a menu. In the same way, in the school where I teach, colleagues often speak of the need to draw by hand, which is irreplaceable to learn to decipher the visible: drawing is learning to see, they explain, that is, learning to decode formal structures of buildings and objects. When drawing, one has to ask certain questions that one would not otherwise ask. Drawing is already thinking. That is why for them, a sketch is an irreplaceable investigative tool. Certainly, as Pierre von Meiss says, "It may seem surprising that in a time of sophisticated photographic and electronic recordings one still insists on the superiority of such an artisanal tool. It's that the observation sketch and its annotations involve an act of selection and human intelligence that have not yet found their equal."[9]

Of course, this idea is not quite new. Jean-Jacques Rousseau was already defending it in 1762 in *Émile*, when he spoke of educating children in drawing:

> I would like [Emile] to cultivate this art, not for the art itself precisely, but to develop a good eye and a flexible hand. ... I will therefore be careful not to provide him with a drawing master, who would only ask him to copy copies and draw from drawings. I want him to have no other master than nature, nor any other models than objects. I want him to have the real thing in front of him and not a copy of it on paper; to draw a house from a house, a tree from a tree, a man from a man; so that he learns to observe them and their appearance.[10]

The virtue of observational drawing—a classic idea that is still very much alive among many professionals today. Thus, Turkle remarks that those who were educated following early CAD practices at the time of Project Athena may be in positions today close enough to their teachers of old, even if, unlike them, they had to face the digital challenge and cannot go back.

But beyond the skepticism around computer rendering, it is interesting to note that the digital also arouses among designers, including contemporary designers, a mistrust of a more philosophical nature. On this point, Japanese designer Kenya Hara, for example, is rather unenthusiastic. In his book *Designing Design*, published in 2007, he writes:

> The remarkable progress of information technology has thrown our society into great turmoil. The computer promises, we believe, to dramatically increase human ability, and the world has overreacted to potential environmental change in that computer-filled future.[11]

According to him, people today are obsessed with the riches computing is supposed to produce; they "do not have time to quietly enjoy the current benefits and already available treasures; and, in leaning so far forward in anticipation of possibilities, they lose their balance."[12] The digital thus appears like a terrain of unbridled acceleration, a whirlwind of technological and stunning blindness, whose meaning is opaque, if not absent, and from which designers should distance themselves in order to hold on to more carnal and more essential things—as if all this contemporary turmoil around the digital were only a kind of slightly gratuitous technophilia, maintained by a handful influential geeks.

As early as the 1990s, however, at the same time the digital elicits concern and mistrust, a number of architects and designers begin to seize the possibilities offered by computers. In architecture, CATIA software, developed and maintained by Dassault Systèmes as early as 1981, has been used ever since 1995 by architects like Frank Gehry, the pioneer of deconstructivism, who uses it to conceive of a good many of his curved and asymmetrical projects, including the famous Guggenheim Museum in Bilbao in 1997. In industrial design, stereolithography opens the door to new object design opportunities. Using a digital model in CAD, this rapid-prototyping technique uses a laser beam to cut, layer, and solidify thin sheets of polymer on each other, which makes it possible to 3D-print unique, solid objects that cannot be produced by any other existing technique. At the moment,

French designer Patrick Jouin is one of those who is exploring this technology with the most finesse and achieving the likes of the One Shot stool, made in one go without any visible supporting shaft or hinge, and that unfolds in one move like an umbrella. Such efforts are still rare and isolated, but they attest concretely to new creative possibilities that the digital offers design.[13] But one cannot yet define a new design discipline: these are simply new working methods that increase possibilities. I propose to bring them together under the term *digitally aided design*. By that, I mean not computer-aided design, which is common in areas as varied as architecture, engineering, physics, and chemistry; I mean it in a more restricted way as computer-aided design specific to the work designers do.

Still, although it is the case that design has used digital tools since the 1980s, *digitally aided design* is not *digital design*. Beyond creative practices alone, "digital technologies have changed the way we interact with things, from the games we play to our work tools," Bill Moggridge writes.[14] In other words, design should take into account not only the digital creative revolution, the one that took place in design and production tools, but also the digital social revolution, the one that intervenes in the use by and lives of people—and that opens an entirely new field: that of digital design strictly speaking. But what is it about?

What in France is called "digital design" comes from what Americans had earlier called "interaction design." As Giuseppe Fioretti and Giancarlo Carbone, engineers at IBM, pointed out, "Interaction design is a user-interface design methodology that was first proposed by Bill Moggridge."[15] Born in 1943, Moggridge is a British industrial designer who was selected in 1979 to design the first laptop, the GRiD Compass, marketed in 1982 and taken on the space shuttle *Discovery* in 1985. For it, he imagined in particular the principle of the flip screen that turns off the computer when you close it. "It was a huge thrill to be a member on a team that was actually doing something so innovative," he says in a video interview.[16] In 1981, the first functional prototype is made, and when he sits down at the screen and starts to interact with the interface, Bill feels completely absorbed in the software. At that moment, he realizes that if he wants to succeed in designing for users, he has to learn how to design interactive technologies. That's when a new design discipline is born: "Designers of products made using digital technologies no longer see their work as designing a (beautiful or useful) physical object, but as a matter of designing interactions with it," says Moggridge. As of then, his agency ID Two becomes one of the first, in the 1980s, to apply methods of industrial design to products with embedded software. With his collaborator Bill Verplank, an MIT engineer specializing in human-machine interfaces who worked on the Xerox Star interface[17] at the end of 1980s, he coins the term *interaction design* to replace *user-interface design*, which computer engineers use.[18] Interaction design, which refers to the design of the interface of computer products, is born.

This new approach is immediately successful. As early as 1989 under Gillian Crampton-Smith's leadership, the Royal College of Art in London creates the first computer-related design department. Faced with the growth of computer technologies that increasingly permeate our lives, the goal of the department, which quickly took on the name "interaction design," is to educate designers capable of creating products and systems that are just as attractive and easy to access as they are useful and functional, all the while searching for ways of imaginatively making the most of digital technologies' possibilities for expression and communication.

Ten years later, in 1999, France steps in line behind the British. Whereas all schools of applied arts are content to approach new technologies under the narrow angle of "multimedia" design, centered since the mid-1980s on CD-ROM design and later expanded to web design, the École nationale supérieure de creation industrielle (Ensci) opens a studio, the Atelier de design numérique (ADN), whose aim is to stand out from the simple "design applied to multimedia," but also, to a lesser extent, from "interaction design" in the American mold. Jean-Louis Frechin, its founder and director for ten years, says, "The concept of digital design was thus born of the need to broaden the concept of interaction and to invent new representations of technical devices with symbolic, aesthetic, functional dimensions focused on uses and people."[19]

At first glance, this does not seem so different from the British notion of "interaction design," which focuses precisely on user-centered design. In the same spirit as the Interaction Design Department of the Royal College of Art in London, the ADN at Ensci seeks to use design to add charm to new technologies, which is far removed from the emptiness Kenya Hara so dreaded: "We are not trying to chase after technologies, but we would like, always deeper in their humanizing effects, to turn on the lights and explore new design territories," continues Jean-Louis Frechin.[20] Nonetheless, he adds, one should understand "digital design" differently from "interaction design": the former would be a more "holistic" and generalist approach that would wed industry and culture as much as the humanities, that is, a more continental approach rooted in the French tradition of design.

If this distinction has its reason for being at a strategic level, it has no theoretical foundation. First, it is not at all certain that the British and the Americans are less concerned than we are about the "humanities." But above all, on the substance, the so-called difference between "interaction design" and "digital design" is not sustainable. Both refer to the same field of application—namely, the field of computer technologies, which is to say computerized products and services emerging from the information technology (IT) sector or from information and communication technologies (ICT), insofar as they involve software and machines capable of executing it: computers. As Richard Stallman, father of free software, aptly points out: "There are many tasks in life for which we use software running on our own machines. This does not just include those devices we call 'personal computers.' It also includes mobile devices, which are also computers. It also includes mobile phones."[21]

Anything that is capable of running software is therefore a computerized product or service, that is, it depends on one or more networked computers and as such is what I call a *digital artifact* in the sense that, as Franck Varenne explains very well, computers are "automata for numerical computation."[22] Therefore, the expressions *interaction design* and *digital design* denote the same thing: both designate design applied to computerized products and services.

And yet the expression *digital design* can be considered more pertinent for two reasons. The first is that it has the merit of referring clearly to the computerized material with which the digital designer must work. Elaborating on an idea of John Maeda, Kenya Hara points out, "The computer is not a tool but a material."[23] Just as clay is a material that the craftsman molds thanks to its infinite plasticity, so too must digital designers understand the characteristics of the material computer they have today in order to know how to exploit its capabilities to produce forms and uses relevant to human beings. Such would be the intent of design (aptly) called *digital*.

The second reason is that the meaning of "interaction design" is broadening more and more beyond the scope of computerized products and services, and moving away from digital material, as evidenced by the official definition of the Interaction Design Association (IxDA):

> Interaction Design defines the structure and behavior of interactive systems. Interaction designers strive to create meaningful relationships between people and the products and services that they use, from computers to mobile devices to appliances and beyond.[24]

Defined thus, interaction design not only covers computerized products and services, but may be everywhere the designer addresses a particular use, that is, the interaction between the user and the product. In the first edition of its *Petit dictionnaire du design numérique* (*Small Dictionary of Digital Design*), the association of French Interactive Designers goes in the same direction and emphasizes that "interaction design is the creation of a dialogue between a person and a product, service, or system" and that this dialogue "may not even be specifically based on the use of an advanced technology";[25] and according to its website, it "makes it easier to make useful, usable, desirable, and profitable products ranging from clock radio to car." By designating as the object of design the interactive move between the user and the product rather than the product alone, "interaction design" becomes synonymous with "user-centered design." Although it may be attractive, this perspective is nevertheless not rigorous (or is, at least, problematic). As I tried to show previously, attention paid to the value of experience, that is, the quality of the lived experience of use, is the fundamental value of any process of design in general: an authentic design is, in essence, an interaction design, that is, centered on the user.

Therefore, it is not legitimate to stretch the meaning of interaction design so broadly. The only valid meaning of the expression "interaction design" can be the one inherited from Bill Moggridge, who coined it to replace "interface design." But what is an interface? The term originally refers to the surface created by the border two bodies have in common. But it's in computer science that the term is mainly used: it refers to a point of intersection between two systems where they can exchange information and therefore communicate, or *interact*. Most often, the systems in question are man and the machine (we speak of human-machine interfaces, HMI), and more specifically, man and computer. This is explained by the fact that computers are the most complex machines ever, but not immediately and directly usable as would be a chair, for example. That's why the computer requires an *interface*, which is to say the means of enabling users to act and react with it in order simply to use it.

What is called a "user interface," then, is a device that determines the material conditions of this interaction between a human being and a machine (HCI); it includes elements that can be both hardware (screen, keyboard, mouse, touchpad, joystick, controller, stylus, clickable scroll wheel) and software (operating systems, programs, web browsers, immersive worlds, online games, mobile apps). There are interfaces in command lines, where users key in instructions to get the computer to execute this or that operation, as well as graphical user interfaces (GUIs), where users are immersed through a video monitor (screen) in a virtual environment (simulation), long modeling a windowed environment or WIMP, based on the metaphor of a desk or desktop. Now, graphic interfaces are undergoing a new revolution, becoming lighter and more mobile, and especially more tactile, thereby involving a new kind of interaction based on finger movements and hand gestures, as with Apple's handheld, multitouch technologies such as the iPhone and iPad, for example.

Thus, where there is no interface, there is no possibility of "interaction design." "Interaction" is the name that we must give the user experience when the experience involves an interface. And an interface is a necessary and essential component of any computerized product or service, which is to say any relying on digital computing automation. On that note, Jean-Louis Frechin is right to point out that "interfaces are an integral part of digital products" and that it is "the result, the purpose, and the aggregation of the process of digital design."[26] In fact, interfaces can be found in any digital artifact, whether a personal computer (PC, Mac, laptop, NetBook), a web server or website, a video game console (PlayStation, Xbox, Wii), a personal digital assistant (Palm), a GPS, an MP3 player, an iPod, a handheld device (iPhone, BlackBerry), a touch pad (iPad, PC Tablet), or a networked object (internet of things).

That's why "digital design" is not and ought not be confused with "digitally aided design." Digitally assisted design produces a result in which there is nothing digital: the digital is only in the process and not in the product. Digital design produces a result entirely or partly made up of digital material: the digital is not only in the process but above all in the product. For example, a book made using InDesign software or QuarkXPress, as are many books printed today, is certainly a digitally produced book, but not a digital book: there is no interface, and none is needed for access. The book is there, here and now, and turns itself over to me in the immediacy of its *phenomenological* physicality, without intermediary and without interactions: I open it, and I can turn the pages with my hands to deliver their content, without needing an interface between them and me. On the other hand, a book available in Apple's iBook on an iPad is certainly a book produced digitally thanks to software engineering, but especially, and above all, it's a digital book: I can access it only by interacting with it (navigating through its content) via an interface and a set of procedures defined by it, which alter the nature and quality of my reading experience.

By "digital design" I mean nothing other than a particular kind of design that one can define thus: creative activity consisting in designing experiences-to-be-lived using interactive shapes produced in computerized products that are organized around an interface. This includes not only the industrial design of computerized products and services (for example, machines, systems, programs), but also specialized disciplines such as game design or web design. By "game design," I mean the process of design inasmuch as it consists of defining rules and gaming mechanisms in order to offer users a video game experience. By "web design," I mean the design process inasmuch as it consists of creating a navigation interface called a "website" in order to offer users the experience of accessing online content. In both cases, there is design solely from the moment that users experience a *design effect*, for example, when a video game provides them with a gaming experience or a website simplifies and improves their experiences accessing content.

Beyond issues of definition, the main problem in digital design today is knowing whether a new design specialization will simply settle alongside those we already are familiar with (product design, space design, graphic design, and so on, to which we would add digital design), or whether we are dealing with a new design paradigm in general that is destined to become part of all existing specializations in order to substantially modify the discipline the way interaction design has. The future will soon tell us if design needs to be fully absorbed in the digital, as some claim, or if it is the digital that must simply be absorbed by design, as I believe.

Postscript
The Design System: Wherein the Author Orders His Principles in Geometric Style

Definitions

I

By "use," I mean the use of a designed object as an *experience-to-be-lived*.

II

By "experience-to-be-lived," I mean a way of acting, feeling, or thinking that results from a *design effect*.

III

By "designed object," I mean *what was the object* of a design process.

IV

By "design process," I mean the creative process aiming at producing a *design effect*.

V

By "design effect," I mean a three-pronged result occurring in a use and transforming it into an experience-to-be-lived.

Axioms

I

A designed object is not necessarily an object (according to definition III).

This has been published as part of "The Effect of Design: A Phenomenological Contribution to the Quiddity of Design Presented in Geometrical Order," trans. Nick Cowling and Marie-Noëlle Dumaz, in *Artifact* 3, no. 4 (2015).

II

The three dimensions of *design effect* are the ontophanic, the callimorphic, and the socioplastic effect (according to chapter 5).

III

A shape can be spatial, volumetric, textile, graphic, or interactive.

Proposition I

Design is the creative activity that consists of designing experiences-to-be-lived by using shapes.

Proof

Design is a process (see definition IV). Its goal is to create experiences-to-be-lived by transforming uses through a design effect (see definitions I, II, and V).

Corollary

The value of the experience is the fundamental value of design.

Proposition II

Design is not the field of objects but the field of effects.

Proof

The purpose of the design process is not to produce objects but to produce effects (see definition IV). A designed object, inasmuch as it is the *object of* a design process (see definition III), may just as well be a material product or an intangible service or a digital interface (according to axiom I). It emerges *as an object*, that is, as an objective phenomenon before a subject only when it releases a design effect that the human subject can experience.

Corollary

A design system is defined by the entirety of the areas of life of shapes seen as fields of application of the design process: architecture, housing, urban space, furniture, clothing, consumer products, print media, interfaces, video games, websites, and so on. That's why, according to the register of forms to which it refers (see axiom III), design can be called "space design," "product design," "industrial design," "fashion design," "graphic design," or "digital design." Despite the diversity of these fields of application, however, design remains one and indivisible. "Whether we create emotion with a spoon or with a space, it's the same to me," says Patrick Jouin.[1]

Scolia

Design is not an area of architecture; it is architecture that is an area of design.

Proposition III

Industrial design is only one area of design.

Proof

The industrial production tool is only a tool. It is not in itself necessary to produce a design effect. It has just been a historical opportunity to invent design, as mechanization technologies emerged in the eighteenth and nineteenth centuries, because these technologies were particularly violent to humans and required greater charm and humanization than any other artifact had had until then. Before the devastation of industrialization, we had no other choice than to invent design, which is to say to delight users with technical objects. Industry was thus the first field of applied design. The need for delight, however, is not unique to the industrial sector. It potentially touches every area of life. That's why design currently reaches beyond the industrial world to which it would be absurd to reduce it. A *design effect*, that is, an experience lived through shapes (according to proposition I), may very well come about with the help of tools other than those of industry, such as public services, the nonprofit sector, the city, the internet, or even crafts.

Corollary

An urban system or a digital service can have more *design effect* than a mass-produced, industrial object.

Scolia

Design is the art of charming everyday life through shapes.

Notes

Preface to the US Edition

1. Samira Ibnelkaïd, "Identité et altérité par écran: modalités de l'intersubjectivité en interaction numérique," PhD diss., Université Lumière, 2016, https:/transphanie.com/.

2. Nathan Jurgenson, "Fear of Screens," *New Inquiry*, January 25, 2016, https://thenewinquiry.com/fear-of-screens/.

3. Nathan Jurgenson, "The IRL Fetish," *New Inquiry*, June 28, 2012. https://thenewinquiry.com/the-irl-fetish/.

4. Pierre Hadot, *Philosophy as a Way of Life: Spiritual Exercises from Socrates to Foucault* (Oxford: Blackwell, 1995).

5. Pieter Vermaas and S. Vial, eds., *Advancements in the Philosophy of Design* (Dordrecht: Springer, 2018).

Part I: Being and the Screen

Foreword to the First Edition

1. Klout is an online service that analyzes user activity on social networks and gives users a score (a "Klout")—a number between one and one thousand supposed to measure their influence online.

2. Evgeny Morozov, *To Save Everything, Click Here: The Folly of Technological Solutionism* (New York: PublicAffairs, 2013).

3. See the philosophical analysis of "order-words" in Gilles Deleuze and Félix Guattari, *A Thousand Plateaus*, trans. Brian Massumi (Minneapolis: University of Minnesota Press, 1987).

4. See Pierre Lévy, *The Semantic Sphere* (Hoboken, NJ: Wiley, 2011).

Introduction

1. See Robert X. Cringely, *The Triumph of the Nerds: The Rise of Accidental Empires*, dir. Paul Sen (Oregon Public Broadcasting, 1996).

2. Bernard Darras, "Machines, complexité et ambition," in *Dessine-moi un pixel: Informatique et arts plastiques*, ed. J. Sultan and B. Tissot (Paris: NPRI/Center Georges-Pompidou, 1991), 107.

3. See "Cinq milliards d'objets connectés," *Le Monde informatique*, August 20, 2010, www.lemondeinformatique.fr/actualites/lire-5-billions-of-the-objects-CONNECTED-summer-31413.html.

4. Sherry Turkle, *Life on the Screen: Identity in the Age of the Internet* (New York: Simon & Schuster, 1995).

5. Pierre Lévy, *Cyberculture*, trans. Robert Bononno (Minneapolis: University of Minnesota Press, 2001), xvi: cyberculture is the "set of technologies (material and intellectual), practices, attitudes, modes of thought, and values that developed along with the growth of cyberspace."

6. Gaston Bachelard, *The New Scientific Spirit*, trans. Arthur Goldhammer (Boston: Beacon Press, 1984), 3.

7. Gilbert Simondon, *On the Mode of Existence of Technical Objects*, trans. Cécile Malaspina and John Rogove (Minneapolis: Univocal Publishing, 2017), 18.

8. We borrow this expression from Jean-Claude Beaune, *L'Automate et ses mobiles* (Paris: Flammarion, 1980), 10.

9. Bernard Darras, "Aesthetics and Semiotics of Digital Design: The Case of Web Interface Design," in *Proceedings of the First INDAF International Symposium Conference* (Incheon, 2009), 11.

10. Daniel Parrochia, "L'Internet et ses représentations," *Philosophies entoilées*, Rue Descartes 55 (2007): 10–20.

11. Peter Sloterdijk, "The Domestication of Being," in *Not Saved: Essays after Heidegger*, trans. Ian Alexander Moore and Christopher Turner (Cambridge: Polity Press, 2017), 96.

12. Bernard Darras and Sarah Belkhamsa, "Do Things Communicate?" *MEI: Mediation Et Information*, nos. 30–31 (2009): 7.

13. We are, of course, referring here to Bachelard's comments at the beginning of "Noumène et microphysique," *Recherches philosophiques* 1 (1931–1932): 55–65.

14. Yann Leroux, "Psychodynamique des groupes sur le réseau Internet" (PhD diss., Université Paris X-Nanterre, 2010), 78, available at the Observatoire Digital Worlds in the Humanities, PDF version.

15. This is Walter Benjamin's expression about art, to which we will return.

Chapter 1

1. Bertrand Gille, *The History of Techniques*, trans. P. Southgate and T. Williamson (Montreux: Gordon and Breach Scientific Publishers, 1986), 10.

2. Gille, *History of Techniques*, 10.

3. Gille, 14.

4. Franck Varenne, *Qu'est-ce que l'informatique?* (Paris: Vrin, 2009), 12.

5. Gille, *History of Techniques*, 14.

6. Gille, 14.

7. A notion we owe to Eric von Hippel, *Democratizing Innovation* (Cambridge, MA: MIT Press, 2005), quoted by Dominique Cardon, "De l'innovation ascendante," InternetActu, June 2005, www.internetactu.net/2005/06/01/de-innovation-up/.

8. Gille, 17 (emphasis in original).

9. Gille, 17.

10. Bertrand Gille, *Engineers of the Renaissance* (Cambridge, MA: MIT Press, 1966).

11. Siegfried Giedion, *Mechanization Takes Command: A Contribution to Anonymous History* (New York: Oxford University Press, 1948).

12. Gille, *History of Techniques*, 17.

13. Jacques Ellul, *The Technological System* (*Le Système technique*) (1977), trans. Joachim Neugroschel (New York: Continuum, 1980), 1. [In what follows, we use the term *technical* to echo Ellul's meaning of *technicien* and *technique*. To distinguish Ellul's discourse from Bertrand Gille's, we use *technology/ical* for *technique* throughout.—Trans.]

14. Ellul, *Technological System*, 1.

15. Ellul, 80.

16. Ellul, 80.

17. Ellul, 80.

18. Immanuel Kant, *Critique of Judgment*, trans. J. H. Bernard (London: Macmillan, 1914), 238, http://lf-oll.s3.amazonaws.com/titles/1217/Kant_0318_EBk_v6.0.pdf.

19. John Boli-Bennet quoted in Ellul, *Technological System*, 82.

20. Jean Baudrillard, *The Consumer Society: Myths and Structures*, trans. Chris Turner (Thousand Oaks, CA: Sage, 1998).

21. Of course, this is not deliberate. There is no anthropomorphism here!

22. Ellul, *Technological System*, 16. The very significant parentheses and exclamation points are Ellul's.

23. Gilbert Simondon, *On the Mode of Existence of Technical Objects*, trans. Cécile Malaspina and John Rogove (Minneapolis: University of Minnesota Press, 2017), 16–17.

24. Martin Heidegger, "The Question Concerning Technology," in *The Question Concerning Technology*, trans. William Lovitt (New York: Harper & Row, 1977), http://www.psyorg/question_concerning_technology.pdf.

25. Herbert Marcuse, *One-Dimensional Man* (Boston: Beacon Press, 1964), xvii.

26. Jürgen Habermas, "Technology and Reification: Technology and Science as 'Ideology,'" trans. Robin Celikates and Rahel Jaeggi, in *The Habermas Handbook* (New York: Columbia University Press, 2018).

27. See Daniel Cérézuelle, "Technique et désir chez Jean Brun," in *Les Philosophes et la technique*, ed. Pascal Chabot and Gilbert Hottois (Paris: Vrin, 2003), 218–220.

28. The formula is from Jean-Pierre Séris.

29. Jean-Pierre Séris, *La Technique* (Paris: Presses universitaires de France, 2000), 377.

30. Séris, *La Technique,* 375.

31. Georges Canguilhem, *The Normal and the Pathological*, trans. Carolyn R. Fawcett (New York: Zone Books, 1991), 33.

32. Simondon, *On the Mode of Existence*, 16.

33. Simondon, 15.

34. Simondon, 18.

35. Simondon, 15.

36. François Dagognet, *Éloge de l'objet* (Paris: Vrin, 1989), 40.

37. Gilbert Hottois, "Les Philosophes et la technique—Les Philosophes et la technique," in Chabot and Hottois, *Les Philosophes et la technique*, 16.

38. Georges Canguilhem, "Machine and Organism," in *The Knowledge of Life*, trans. Stefanos Geroulanos and Daniela Ginsburg (New York: Fordham University Press, 2008), 93.

39. Séris, *La Technique*, 268.

40. Henry Cole, quoted by Alexandra Midal, *Design: Introduction à l'histoire de la discipline* (Paris: Pocket, 2009), 33–34.

41. The expression is from Jacques Viénot. In Étienne Souriau, *Vocabulaire d'esthétique* (Paris: Presses universitaires de France, 2004), 880.

42. See *A Short Treatise on Design* in this book.

43. Séris, *La Technique*, 266.

44. Séris, 267.

45. Bernard Stiegler, *De la misère symbolique*, vol. 1: *L'Époque hyperindustrielle* (Paris: Galilée, 2004).

46. Séris, *La Technique*, 280.

47. Steve Jobs, interview for CNN Tech, June 2011, quoted by George Beahm, *I, Steve: Steve Jobs in His Own Words* (Chicago: B2 Books, 2011), 30.

48. Steve Jobs, iPad presentation speech, 2010, quoted by Steven Johnson, "Marrying Tech and Art," *Wall Street Journal*, August 27, 2011, http://online.wsj.com/article/SB10001424053111904875404576532342684923826.html/.

49. Steve Jobs, in Robert X. Cringely, *The Triumph of the Nerds: The Rise of Accidental Empires*, dir. Paul Sen (Oregon Public Broadcasting, 1996), 28'50 https://www.youtube.com/watch?v=sX5g0kidk3Y.

50. See Walter Isaacson, *Steve Jobs* (New York: Simon & Schuster, 2011).

51. Henri Bergson, *Time and Free Will: Essay on the Immediate Data of Consciousness*, trans. F. L. Pogson (London: George Allen & Unwin, 1921), 172.

52. Séris, *La Technique*, 383.

53. Lawrence Lessig, *Free Software, Free Society: Selected Essays of Richard M. Stallman* (Boston: Gnu Press, 2002), 11, http://www.gnu.org/philosophy/lessig-fsfsintro.html.

54. Friedrich Nietzsche, "On the Uses and Disadvantages of History for Life," trans. Ian Johnstone, 16, http://www.leudar.com/library/On%20the%20Use%20and%20Abuse%20of%20History.pdf.

55. Friedrich Nietzsche, *Twilight of the Idols*, in *The Portable Nietzsche*, ed. and trans. Walter Kaufmann (New York: Penguin Books, 1976).

56. Steve Jobs, interview in *Playboy Magazine* (February 1985), quoted by Beahm, *I, Steve*, 78.

57. Séris, *La Technique*, 2–3.

58. Pierre Musso, Laurent Ponthou, and Eric Seulliet, *Fabriquer le futur, 2: L'Imaginaire au service de l'innovation* (Paris: Pearson Education France, 2007), 6.

59. Séris, *La Technique*, 3.

60. Séris, 5.

61. Séris, 6.

62. Dagognet, *Éloge de l'objet*, 18–184.

63. Séris, *La Technique*, 58.

64. Séris, 59.

65. Séris, 50–51.

Chapter 2

1. Thomas S. Kuhn, *The Structure of Scientific Revolutions* (Chicago: University of Chicago Press, 1962).

2. Siegfried Giedion, *Mechanization Takes Command: A Contribution to Anonymous History* (Oxford: Oxford University Press, 1948).

3. Bertrand Gille, *The History of Techniques*, trans. P. Southgate and T. Williamson (Montreux: Gordon and Breach Scientific Publishers, 1986), 671.

4. Michel Serres, "Ce n'est pas une crise, c'est un changement de monde," *Le Journal du dimanche*, December 30, 2012, https://www.lejdd.fr/Economie/Serres-Ce-n-est-pas-une-crise-c-est-un-changement-de-monde-583645-3134546.

5. Gille, *History of Techniques*, 745.

6. Jean-Pierre Séris, *La Technique* (Paris: Presses universitaires de France, 2000), 65.

7. Gille, *History of Techniques*, 752.

8. Gille, 745.

9. Gille, 745–746.

10. Gille, 752, 787–793, 796–804.

11. Jacques Ellul, *The Technological System*, trans. Joachim Neugroschel (New York: Continuum, 1980), 93.

12. Gille, *History of Techniques*, 796.

13. Ellul, *Technological System*, 98.

14. Ellul, 98.

15. Philippe Breton, *The Culture of the Internet and the Internet as Cult: Social Fears and Religious Fantasies*, trans. David Bade (Duluth, MN: Litwin Books, 2011), 1.

16. Jeremy Rifkin, *The Third Industrial Revolution: How Lateral Power Is Transforming Energy, the Economy, and the World* (New York: Palgrave Macmillan, 2011), 1.

17. Rifkin, *Third Industrial Revolution*, 2, 5.

18. Michel Volle, *L'Économie des nouvelles technologies* (Paris: Economica, 1999), 3.

19. Steve Jobs, in Julian Krainin and Michael R. Lawrence, *Memory and Imagination: New Pathways to the Library of Congress* (Baltimore: Michael Lawrence Films and Krainin Productions, 1990), TV documentary.

20. Gille, *History of Techniques*, 820.

21. Volle, *L'Économie des nouvelles technologies*, 3.

22. Volle, 26.

23. Volle, 4.

24. Volle, 26.

25. Sylvie Leleu-Merviel, "Les désarrois des 'Maîtres du sens' à l'ère du numérique," in *Hypertextes, hypermedia: Créer du sens à l'ère du numérique, H2PTM'03*, Lavoisier, 2003, 19.

26. Volle, *L'Économie des nouvelles technologies*, 26.

27. Volle, 3.

28. Volle, 26.

29. Volle, 26.

30. Volle, 24.

31. Hubert Guillaud, "Où va l'économie numérique? (1/3): Vers une innovation sans emplois?" InternetActu.net, February 1, 2012, http://www.internetactu.net/2012/02/01/ou-va-leconomie-numerique-13-vers-une-Innovation-sans-emplois/.

32. Volle, *L'Économie des nouvelles technologies*, 26 (emphasis in the original).

33. Volle, 32.

34. Giedion, *Mechanization Takes Command*.

35. Michel Volle, *De l'informatique: Savoir vivre avec l'automate* (Paris: Economica, 2006), 3, http://www.volle.com/ouvrages/informatique/informatique1.pdf.

36. Michel Volle, "Comprendre la croissance à l'heure de l'informatisation de la société," in *Une Croissance intelligente*, ed. Philippe Lemoine (Paris: Descartes & Cie, 2012),

published on InternetActu.net under the title "Comprendre la croissance à l'heure de l'informatisation de la société," April 27, 2012, www.internetactu.net/2012/04/27/comprendre-la-croissance-a-lheure-de-linformatisation-de-la-societe/.

37. Volle, "Comprendre la croissance à l'heure."

38. Volle.

39. Volle.

40. Jean-Claude Beaune was already talking about "total technological fact" in 1980 when discussing robots. See his *L'Automate et ses mobiles* (Paris: Flammarion, 1980), 7.

41. Pierre Lévy, *Cyberculture*, trans. Robert Bononno (Minneapolis: University of Minnesota Press, 2001), xvi.

42. Paul Mathias, *Qu'est-ce que l'Internet?* (Paris: Vrin, 2009), 61.

43. Mathias, *Qu'est-ce que l'Internet?* 31.

44. Volle, *De l'informatique*, 3.

45. Volle, "Comprendre la croissance."

Chapter 3

1. Anne Cauquelin, *L'Invention du paysage* (Paris: Presses Universitaires de France, 2000), 35.

2. Here we are referring to the hyperbolic geometry of N. I. Lobatchevsky or to G. F. B. Riemann's elliptical geometry, which put into question Euclid's fifth postulate, according to which, "in a plane, given a line and a point not on it, at most one line parallel to the given line can be drawn through the point" (https://en.wikipedia.org/wiki/Playfair%27s_axiom).

3. Claude Lévi-Strauss, *Entretiens avec Georges Charbonnier* (Paris: Union générale d'éditions, 1998).

4. Jean-Michel Besnier, *Les Théories de la connaissance* (Paris: Presses universitaires de France, 2011), 116.

5. Jean-Louis Le Moigne, *Les Épistémologies constructivistes* (Paris: Presses universitaires de France, 2007), 61.

6. Jean Piaget, quoted by Le Moigne, *Les Épistémologies constructiviste*, 75.

7. Gaston Bachelard, *The Formation of the Scientific Mind*, trans. and anno. Mary McAllester Jones (Manchester: Clinamen, 2002), 25.

8. Gaston Bachelard, "Noumenon et microphysique," *Recherches philosophiques* 1 (1931–1932), 11–22.

9. Bachelard, *Formation of the Scientific Mind*, 69, 70 (emphasis in the original).

10. Gaston Bachelard, *Atomistic Intuitions: An Essay on Classification*, trans. Roch C. Smith (Albany, NY: SUNY Press, 2018), 90.

11. Bachelard, *Formation of the Scientific Mind*, 240.

12. Gaston Bachelard, *The New Scientific Spirit,* trans. Arthur Goldhammer (Boston: Beacon Press, 1984), 13.

13. Bachelard, *Formation of the Scientific Mind*, 216.

14. Gaston Bachelard, *L'Activité rationaliste de la physique contemporaine* (Paris: Presses universitaires de France, 1965), 5.

15. Gaston Bachelard, *Études* (Paris: Vrin, 2002), 17.

16. Bachelard, *Études*, 16.

17. Sheila Jones, *The Quantum Ten: A Story of Passion, Tragedy, Ambition, and Science* (New York: Oxford University Press, 2008).

18. Bachelard, *Études*, 18, 22.

19. Bachelard, *L'Activité rationaliste*, 10.

20. Pierre Lévy, *Cyberculture*, trans. Robert Bononno (Minneapolis: University of Minnesota Press, 2001), 4, 5:

> Is technology an autonomous factor, separate from society and culture, which are not more than passive entities pierced by some external agent? It is my contention that technology is a way of analyzing global sociotechnical systems, a point of view that emphasizes the material and artificial components of human phenomena, and not a real entity, which exists independently, has distinct effects, and acts on its own. ... There are therefore no genuine relationships between "a" technology (part of the cause) and "a" culture (which would undergo its effects), but among a multitude of human agents who variously invent, produce, use, and interpret *technologies*.

21. Mircea Éliade first used the term *ontophany* in 1956 in the sense of the "manifestation of a being" in *The Sacred and the Profane: The Nature of Religion* (New York: Harper & Row, 1961), 97, 117. It is quoted twice alongside the term *hierophany*, which means "the act of manifestation of the sacred," which is the central concept of his book. According to Éliade, for a religious man, the modalities of manifestation of being and of sacredness are inseparable: "Ontophany and hierophany come together," he writes. The term is reused in a similar sense by Vladimir Jankélévitch in 1957 in *Le Je-ne-sais-quoi and le presque-rien*, vol. 1: *La Manière et l'occasion* (Paris: Le Seuil, 1980), 34.

22. Not to be confused with the notion of "technophany" that Simondon developed following Éliade's "hierophany." Cf. Gilbert Simondon, "Psycho-sociologie de la technicité," *Bulletin de l'École pratique de psychologie et de pédagogie*, no. 2 (1960): 127–140; no. 3 (1961), 227–238; no. 4 (1961), 319–350.

23. This is Lewis Mumford's term to describe the combination of wood and water in the Renaissance. See L. Mumford, *Technics and Civilization* (New York: Harcourt & Brace, 1934).

24. This point, however, deserves a discussion to determine whether the technical conditioning of perception in the theory of ontophany is similar or not to Peter-Paul Verbeek's theory of technical mediation.

25. Cauquelin, *L'Invention du paysage*, 31. See also the landmark essay by Alain Roger, *Court traité du paysage* (Paris: Gallimard, 1997).

26. Oscar Wilde, "The Decay of Lying," in his *Intentions* (New York: Brentano's, 1905), 41.

27. Pierre Lévy, *La Machine univers: Création, cognition et culture informatique* (Paris: La Découverte, 1987), 213.

28. Bell filed his patent on February 14, 1876, at 2:00 p.m., and Gray filed his the same day at 4:00 p.m. "Au premier la gloire, au second l'oubli" ("To the first, glory; to the second, oblivion"), as Robert Vignola points out in his book *Allô! La merveilleuse aventure du téléphone* (Paris: CZ Créations, 2000), 18.

29. Walter Benjamin, "A Short History of Photography," trans. Stanley Mitchell, *Screen*, March 1, 1972, 5–26, https://doi.org/10.1093/screen/13.1.5.

30. On the complex editorial history of this book, see the "Notice" of the French translation of W. Benjamin's article in *L'Œuvre d'art à l'époque de sa reproductibilité technique*, trans. Lionel Dovoy (Paris: Allia, 2009), 79.

31. Benjamin, "A Short History of Photography," 6.

32. Benjamin, 6.

33. Benjamin, 25.

34. Benjamin, 25. The same definition and the same example are repeated in Walter Benjamin, "The Work of Art in the Age of Its Technological Reproducibility," *Grey Room* 39 (2010): 15, https://www.mitpressjournals.org/doi/abs/10.1162/grey.2010.1.39.11.

35. Benjamin, "The Work of Art," 27.

36. Benjamin, "A Short History of Photography," 23.

37. Benjamin, "The Work of Art," 15.

38. Benjamin, "A Short Story of Photography," 19.

39. Benjamin, "A Short Story of Photography," 7.

40. Benjamin, "The Work of Art," 14.

41. Benjamin, "The Work of Art," 30–31.

42. Benjamin had already pointed out on another page: "Just as the entire mode of existence of human collectives changes over long historical periods, so too does their mode of perception. The way in which human perception is organized—the medium in which it occurs—is conditioned not only by nature but by history." Benjamin, "The Work of Art," 23.

43. "For several years, the requirements of theoretical work for me have come from my painting practice." Pierre-Damien Huyghe, *Du commun: Philosophie pour la peinture et le cinéma* (Belval: Circé, 2002), 9.

44. Huyghe, *Du commun*, 82.

45. Huyghe, 64.

46. Huyghe, 113.

47. Huyghe, 96. The same idea emerges even more prominently a few years later: "Even if 'art' can be a method of meaning, this method discovered its philosophical foundation first in the technological." P.-D. Huyghe, *Le Différend esthétique* (Belval: Circé, 2004), 11.

48. Huyghe, *Le Différend esthétique*, 114.

49. Huyghe, *Le Différend esthétique*, 114.

50. Pierre-Damien Huyghe, "Introduction au dossier 'Temps et appareils,'" *Plastik*, no. 3 (2003): 4.

51. Huyghe, "Introduction," 4.

52. Huyghe, 4.

53. Huyghe, 4.

54. Huyghe, 5.

55. Huyghe, 5.

56. Huyghe, *Le Différend esthétique*, 110–111.

57. Pierre-Damien Huyghe (ed.), *L'Art au temps des appareils* (Paris: L'Harmattan, 2005), 11.

58. Pierre-Damien Huyghe, *Modernes sans modernité* (Paris: Nouvelles Éditions Lignes, 2009), 113.

59. Huyghe, *Modernes sans modernité*, 111.

60. Huyghe, *L'Art au temps*, 25–26.

61. Huyghe, *Modernes sans modernité*, 120.

62. Huyghe, *L'Art au temps*, 25–26.

63. Vignola, *Allô!*

64. Vignola, 22.

65. Vignola, 24.

66. Vignola, 24.

67. Pauline de Broglie, *Comment j'ai vu 1900* (Paris: Grasset, 1962) (emphasis added).

68. Herbert N. Casson, *The History of the Telephone* (Chicago: McClurg, 1910), chap. 6, first sentence, http://etext.lib.virginia.edu/toc/modeng/public/CasTele.html.

69. Serge Tisseron, *Virtuel, mon amour: Penser, aimer, souffrir à l'ère des nouvelles technologies* (Paris: Albin Michel, 2008), introduction.

70. Sherry Turkle, *Alone Together: Why We Expect More from Technology and Less from Each Other* (New York: Basic Books, 2012).

71. Michel Serres, *Thumbelina: The Culture and Technology of Millennials*, trans. Daniel W. Smith (London: Rowman & Littlefield International, 2014).

Chapter 4

1. See Serge Proulx and Guillaume Latzko-Toth, "La Virtualité comme catégorie pour penser le social: L'usage de la notion de communauté virtuelle," *Sociologie et Société* 32, no. 2 (2000): 99–122.

2. Aristotle, *The Metaphysics*, trans. Hugh Tredennick (Cambridge, MA: Harvard University Press, Loeb Library, 1933–1935), Book IX, VI, l. 3, 447.

3. Aristotle, *Metaphysics*, 445.

4. Gilles-Gaston Granger, *Le Probable, le possible et le virtuel: Essai sur le rôle du non-actuel dans la pensée objective* (Paris: Odile Jacob, 1995), 13.

5. A. Bertrand, "Virtuel," in *Les Notions philosophiques: Encyclopédie philosophique universelle*, vol. 2 (Paris: Presses universitaires de France, 1990), 2745.

6. André Lalande, "Virtuel," in *Vocabulaire technique et critique de la philosophie* (Paris: Presses universitaires de France, 1960), 1211–1212.

7. Pierre Quéau, *Le Virtuel: Vertus et vertige* (Seyssel: Champ Vallon, Milieux, 1986), 26.

8. Quéau, *Le Virtuel*, 27.

9. See https://en.wikipedia.org/wiki/Virtual_memory.

10. See https://en.wikipedia.org/wiki/Virtual_memory.

11. Jean-François Bach, Olivier Houdé, Pierre Lena, and Serge Tisseron, *L'Enfant et les écrans: Un Avis de l'Académie des sciences* (Paris: Le Pommier, 2013), sec. 5.1, 19, https://www.enssib.fr/bibliotheque-numerique/documents/60271-l-enfant-et-les-ecrans.pdf.

12. Bach et al., *L'Enfant et les écrans*, sec. 5.1.2, 20.

13. Stéphane Vial, "Il était une fois 'pp7,' ou la naissance d'un groupe sur l'Internet: Retour sur la socialisation en ligne d'une communauté étudiante," *Réseaux* 6, no. 164 (2010): 65.

14. "According to wikipedia.fr, the term *virtual* is used to "describe what is happening in a computer or on the internet, that is to say in a 'digital world' as opposed to the 'physical world.'" Wikipedia, s.v. "Virtuel," last modified, November 21, 2018, http://fr.wikipedia.org/wiki/Virtuel.

15. Serge Tisseron, *Rêver, fantasmer, virtualiser: Du virtuel psychique au virtuel numérique* (Paris: Dunod, 2012).

16. Nicolas Nova, "Famous User Figures in the History of HCI," February 18, 2010, http://www.nicolasnova.net/pasta-and-vinegar/2010/02/18/famous-user-figures-in-the-history-of-hci.

17. Bernard Darras, "Machines, complexité et ambition," in *Dessine-moi un pixel: Informatique et arts plastique*, ed. J. Sultan and B. Tissor (Paris: INRP/Centre Georges-Pompidou, 1991), 107.

18. Darras, "Machines," 107.

19. Susan Kare, User Interface Graphics, http://kare.com.

20. Robert X. Cringely, *The Triumph of the Nerds: The Rise of Accidental Empires*, dir. Paul Sen (Oregon Public Broadcasting, 1996).

21. This is the principle of the WIMP (windows, icons, menus, pointing device) interface, invented by Xerox in the 1970s, improved and marketed by Apple in the 1980s, and imposed on everyone by Microsoft in the 1990s.

22. Sherry Turkle, *Life on the Screen: Identity in the Age of the Internet* (New York: Simon & Schuster, 1995), 19.

23. Turkle, *Life on the Screen*, 19.

24. Philippe Quéau, *Éloge de la simulation: De la vie des langages à la synthèse des images* (Seyssel: Champ Vallon, 1986).

25. Quéau, *Le Virtuel*.

26. Quéau, 13.

27. Quéau, 14.

28. Quéau, 9.

29. Quéau, 13–14.

30. Quéau, 16.

31. *Second Life*, http://secondlife.com.

32. Quéau, "La Pensée virtuelle," *Réseaux*, no. 61 (September–October 1993): 69.

33. Quéau, *Metaxu: Théorie de l'art intermédiaire* (Seyssel: Champ Vallon/INA, 1989).

34. Quéau, *Le Virtuel*, 18.

35. Quéau, *Le Virtuel*, 45.

36. Quéau, "La Pensée virtuelle," 71.

37. Quéau, "La Pensée virtuelle," 72.

38. Quéau, "La Pensée virtuelle," 71.

39. Pierre Lévy, *Qu'est-ce que le virtuel?* (Paris: La Découverte, 1998).

40. Gaston Bachelard, *The Psychoanalysis of Fire*, trans. Alan C. M. Ross (Boston: Beacon Press, 1964), 3.

41. Bachelard, *Psychoanalysis of Fire*, 1.

42. Bachelard, 3.

43. Bachelard, 4.

44. For example, Pierre Lévy, *Qu'est-ce que le virtuel?* or Alain Milon, *La Réalité virtuelle: Avec ou sans le corps?* (Paris: Autrement, 2005).

45. Bachelard, *Psychoanalysis of Fire*, 4.

46. Turkle, *Life on the Screen*, 23.

47. Turkle, 23–24.

48. Turkle, 9.

49. William Gibson, *Neuromancer* (New York: Ace Books, 1984).

50. See our lecture slides, "There Is No Difference between the 'Real' and the 'Virtual': A Brief Phenomenology of the Digital Revolution," speech at Theorizing the Web 2013, Graduate Center, City University of New York, March 2, 2013, http://goo.gl/qhUOJ.

Chapter 5

1. Wikipedia, s.v. "Virtuel," last modified November 21, 2018, http://fr.wikipedia.org/wiki/Virtuel.

2. Gaston Bachelard, *Études* (Paris: Vrin, 2002), 17.

3. Gilles-Gaston Granger, *Le Probable, le possible et le virtuel: Essai sur le rôle du non-actuel dans la pensée objective* (Paris: Odile Jacob, 1995).

4. Granger, *Le Probable*, 11.

5. Granger, 13.

6. Granger, 13.

7. Granger, 14.

8. Granger, 15.

9. Granger, 17.

10. Granger, 80.

11. Granger, 9.

12. Granger, 16.

13. Granger, 15.

14. Granger, 80.

15. Granger, 81.

16. Granger, 81.

17. Granger, 99.

18. Granger, 129.

19. Granger, 231.

20. Granger, 231.

21. Granger, 232.

22. Granger, 234.

23. See Stéphane Vial, "Qu'appelle-t-on design numérique?" *Interfaces numériques* 1, no. 1 (2012): 91–106.

24. Pierre Lévy, "Le Médium algorithmique," *Sociétés* 3, no. 129 (2015): 79–96, http://pierrelevyblog.files.wordpress.com/2013/02/00-le_medium_algorithmique.pdf.

25. Paul Mathias, *Qu'est-ce que l'Internet?* (Paris: Vrin, 2009), 32.

26. Philippe Breton, *Une Histoire de l'informatique* (Paris: La Découverte, 1987), 164.

27. Breton, *Une Histoire de l'informatique*, 180.

28. See Sherry Turkle, *Simulation and Its Discontents* (Cambridge, MA: MIT Press, 2009).

29. Quéau, *Le Virtuel: Vertus et vertiges* (Seyssel: Champ Vallon, "Milieux" 1993), 30.

30. Quéau, *Le Virtuel*, 29.

31. Quéau, 30.

32. Quéau, 45.

33. Mathias, *Qu'est-ce que l'Internet?* 55.

34. Quéau, *Le Virtuel*, 15.

35. Lev Manovich, *Software Takes Command* (New York: Bloomsbury, 2013). This title refers to that of S. Giedion's *Mechanization Takes Command: A Contribution to Anonymous History* (Oxford: Oxford University Press, 1984).

36. See my blog post, "La Menace très fantomatique du 'trading algorithmique,'" *Reduplikation.net*, May 20, 2013, http://reduplikation.tumblr.com/post/50921636485/la-menace-tr%C3%A8s-fantomatique-du-trading.

37. Sylvie Leleu-Merviel, "Les Désarrois des 'Maîtres du sens' à l'ère du numérique," *Hypertextes, hypermédias: Créer du sens à l'ère numérique, H2PTM'03* (Lavoisier 2003), 20, http://halshs.archives-ouvertes.fr/hal-00467743.

38. Pierre Lévy, *De la programmation comme l'un des beaux-arts* (Paris: La Découverte, 1992).

39. Lévy, *De la programmation*, 7.

40. Robert X. Cringely, *The Triumph of the Nerds: The Rise of Accidental Empires*, dir. Paul Sen (Oregon Public Broadcasting, 1996).

41. The first successful minicomputer, launched commercially in 1965.

42. Steve Wozniak in Cringely, *Triumph of the Nerds*.

43. Wozniak in Cringely, *Triumph of the Nerds*.

44. Bill Moggridge, *Designing Interactions* (Cambridge, MA: MIT Press, 2007), http://www.designinginteractions.com/chapters/introduction.

45. Moggridge, *Designing Interactions*.

46. Sherry Turkle, *Life on the Screen: Identity in the Age of the Internet* (New York: Simon & Schuster, 1995).

47. Steve Jobs, in Cringely, *Triumph of the Nerds,* https://www.pbs.org/nerds/part1.html.

48. Mathieu Triclot, *Philosophie des jeux vidéos* (Paris: La Découverte, 2011), 21.

49. Triclot, *Philosophie des jeux vidéos,* 96.

50. L. Floridi, "The Unsustainable Fragility of the Digital," *Philosophy & Technology* 30, no. 3 (September 2017): 259–261.

51. Sextus Empiricus, *Esquisses pyrrhoniennes* (Paris, Le Seuil, 1997), I, 71.

52. Daniel Parrochia (dir.), *Penser les réseaux* (Seyssel: Champ Vallon, 2001).

53. Pierre Musso, *Critique des réseaux* (Paris: Presses universitaires de France, 2003).

54. Founded in 1983, the magazine *Reseaux* addresses the whole field of communications with a particular focus on telecommunications. See http://www.cairn.info/revue-reseaux.htm.

55. Mathias, *Qu'est-ce que l'Internet,* 25.

56. A. A. Casilli, *Les Liaisons numériques: Vers une nouvelle sociabilité?* (Paris: Le Seuil, 2010), 8.

57. S. Lebovici, quoted by Sylvain Missonnier, "Une Relation d'objet virtuel?" *Le Carnet psy,* no. 120, (2007): 43–47, http://www.cairn.info/revue-le-carnet-psy-2007-7-page-43.htm.

58. Casilli, *Les Liaisons numériques,* 229–230.

59. Sherry Turkle, *Alone Together: Why We Expect More from Technology and Less from Each Other* (New York: Basic Books, 2012).

60. Casilli, *Les Liaisons numériques,* 325.

61. Steven Levy, *The Perfect Thing: How the iPod Shuffles Trade, Culture and Coolness* (New York: Simon & Schuster, 2006).

62. William Shakespeare, *Hamlet,* act 1, scene 2.

63. Heraclitus, *Fragments,* B I.

64. Triclot, *Philosophie des jeux vidéos,* 21.

65. Triclot, 22.

66. John Thackara, *In the Bubble: Designing in a Complex World* (Cambridge, MA: MIT Press, 2005), 11.

67. Antoine-Laurent Lavoisier, *Elements of Chemistry* (1740), trans. Robert Kerr, http://www.gutenberg.org/cache/epub/30775/pg30775.txt.

68. Jean-Pierre Séris, *La Technique* (Paris: Presses universitaires de France, 2000), 378.

69. Yann Leroux, "En lisant Stéphane Vial," *Mondes numériques*, May 8, 2009, https://groups.google.com/forum/#!topic/mondes-numeriques/juF4MSshTg.

70. Quéau, *Le Virtuel*, 42.

71. Stéphane Vial, "Il était une fois 'pp7,' ou la naissance d'un groupe sur l'Internet: Retour sur la socialisation en ligne d'une communauté étudiante," *Réseaux* 164 (2010): 64.

72. Casilli, *Les Liaisons numériques*, 123.

73. Vial, "Il était une fois," 64.

74. Quéau, *Le Virtue*, 15.

75. See Triclot, *Philosophie des jeux vidéos*.

76. Triclot, *Philosophie des jeux vidéos*, 231.

77. Triclot, 234.

78. Sébastien Genvo, "Penser les phénomènes de 'ludicisation' du numérique: Pour une théorique de la jouabilité," *Revue des sciences sociales*, no. 45 (2011): 69.

79. Genvo, "Penser les phénomènes de 'ludicisation' du numérique."

80. See Triclot, *Philosophie des jeux vidéos*, 24, and Genvo, "Penser les phénomènes de 'ludicisation' du numérique," 70–71.

81. *Hegel's Aesthetics*, ed. and trans. T. M. Knox, volume 6 (Oxford: Oxford University Press, 2015), sec. 1, 31.

82. Bernard Darras, "Ambition et création artistique assistée par ordinateur," in *Faire/voir et savoir: Connaissance de l'image, image et connaissance*, ed. Brigitte Poirier and Josette Sultan (Paris: NPRI, 1992), 89.

83. Sigmund Freud, *On Creativity and the Unconscious* (New York: Harper Perennial Modern Classics, 2009).

84. Genvo, "Penser les phénomènes de 'ludicisation' du numérique," 72.

Chapter 6

1. Peter Sloterdijk, "Foreword to *The Theory of Spheres*," in *Cosmograms*, ed. Melik Ohanian and Jean-Christophe Royoux (New York: Lukas & Sternberg, 2005), 232.

2. Sloterdijk, "Foreword," 231.

3. Bruno Latour, "A Cautious Prometheus? A Few Steps toward a Philosophy of Design (with Special Attention to Peter Sloterdijk)," in *Proceedings of the 2008 Annual*

International Conference of the Design History Society, ed. F. Hackne, J. Glynne, and V. Minto (Universal Publishers, 2009), 8. E-book.

4. Mathieu Triclot, *Philosophie des jeux vidéos* (Paris: La Découverte, 2011), 16.

5. A. Beyaert-Geslin, "Formes de table, formes de vie: Réflexions sémiotiques pour vivre ensemble," *MEI: Médiation et Information*, nos. 30–31 (2009): 99–110.

6. Louis H. Sullivan, "The Tall Office Building Artistically Considered," *Lippincott's Magazine* 57 (March 1896).

7. Jean Baudrillard, *The Consumer Society: Myths and Structures*, trans. Chris Turner (Thousand Oaks, CA: Sage, 1998).

8. Beyaert-Geslin, "Formes de tables," 100.

9. Beyaert-Geslin, 102.

10. Beyaert-Geslin, 100.

11. A. Findeli, "Searching for Design Research Questions: Some Conceptual Clarifications," in *Questions, Hypotheses, and Conjectures: Discussions on Projects by Early Stage and Senior Designers*, ed. R. Chow, G. Joost, and W. Jonas (Bloomington, IN: iUniverse, 2010), 292.

12. Stéphane Vial, *Short Treatise on Design*, this volume, 181.

13. Vial, this volume, 174–175.

14. Vial, this volume, 198.

15. See the commentary given by Mads Nygaard Folkmann in *The Aesthetics of Imagination in Design* (Cambridge, MA: MIT Press, 2013), 188–189.

16. Vial, *Short Treatise on Design*, this volume, 173–178.

17. See also Stéphane Vial, "Le Geste du design et son effet: Vers une philosophie du design," *Figures de l'art: revue d'études esthétiques*, no. 26 (2013).

18. Beyaert-Geslin, "Formes de tables," 107.

19. Ronan Bouroullec on Joyn: "Our grandparents lived on a farm where the kitchen was the focus of the house. This table was the place where people ate, where they talked, and I'm sure my father did his homework there. It was just a surface." Interview with Lucia Allais, Ronan and Erwan Bouroullec, Paris, Phaidon, 2003, http://www.bouroullec.com/.

20. Beyaert-Geslin, "Formes de tables," 108.

21. Beyaert-Geslin, 109.

22. See Stéphane Vial, "Qu'est-ce que le design numérique?" *Interfaces numériques* 1, no. 1 (2012), 91–106.

23. Mathieu Triclot, *Philosophie des jeux vidéos* (Paris: La Découverte, 2011), 15–16.

24. Triclot, 16.

25. Triclot, 14.

26. Triclot, 15.

27. Triclot, 21.

28. Bernard Darras, "Aesthetics and Semiotics of Digital Design: The Case of Web Interface Design," in *Proceedings of the INDAF International Conference* (Incheon, Korea, 2009), 13.

29. "Google lance la voiture sans chauffeur," *Le Monde.fr*, May 9, 2012, www.lemonde.fr/technologies/video/2012/05/09/googlelance-the-car-in-chauffeur_1698400_651865.html.

30. Sherry Turkle, *Simulation and Its Discontents* (Cambridge, MA: MIT Press, 2009), 6.

31. Nathan Jurgenson, *The New Inquiry*, November 13, 2013.

32. Paul Miller, "I'm Still Here: Back Online after a Year without the Internet," *Verge*, May 1, 2013, www.theverge.com/2013/5/1/4279674/im-still-here-back-online-after-a-year-without-the-internet.

33. Jean-Claude Beaune, *La Technologie* (Paris: Presses universitaires de France, 1972), 5.

34. Les Éditions Volumiques, whose creative director is Étienne Mineur, is a digital design studio based in Paris that invents new types of games and toys based on the implementation of the relationship between the tangible and the digital. See http://volumique.com.

Conclusion

1. Bruno Latour, *Aramis, or the Love of Technology*, trans. Catherine Porter (Cambridge, MA: Harvard University Press, 1996), viii.

2. Mathieu Triclot, *Philosophie des jeux vidéos* (Paris: La Découverte, 2011), 16.

3. Bruno Latour, *We Have Never Been Modern*, trans. Catherine Porter (Cambridge, MA: Harvard University Press, 1993).

4. Bruno Latour, "A Cautious Prometheus? A Few Steps toward a Philosophy of Design (with Special Attention to Peter Sloterdijk)," in *Proceedings of the 2008 Annual International Conference of the Design History Society*, ed. F. Hackne, J. Glynne, and V. Minto (Universal Publishers, 2009), 2–10.

5. Anne Cauquelin, *L'Invention du paysage* (Paris: Presses universitaires de France, 2000).

6. Peter Sloterdijk, "Foreword to the Theory of Spheres" in *Cosmograms*, edited by Melik Ohanian and Jean-Christophe Royoux (New York: Lukas & Sternberg, 2005), 230.

7. Geneviève Lombard, "Le Non-virtualisable de la psychanalyse," *Inconscient.net*, Bordeaux, September 26, 2007, http://inconscient.net/non_virtualisable.htm.

Supplement 1

1. Sherry Turkle, *Life on the Screen: Identity in the Age of the Internet* (New York: Simon & Schuster, 1995), 9.

2. On this subject, see Frédéric Joignot's excellent article, "L'Amitié à l'épreuve de Facebook," *Le Monde*, January 2, 2014, https://www.lemonde.fr/culture/article/2014/01/02/l-amitie-a-l-epreuve-de-facebook_4342222_3246.html.

3. Aristotle, *Nicomachean Ethics*, trans. H. Rackham (Cambridge, MA: Harvard University Press, Loeb Classics, 1934), book 8, 155a, 24.

4. Nathan J. Jurgenson and P. J. Rey, "About Cyborgology," *The Society Pages—Cyborgology*, University of Minnesota, http://thesocietypages.org/cyborgology/about.

Supplement 2

1. Daniel C. Dennett, *Consciousness Explained* (Boston: Little, Brown, 1991).

2. Stéphane Leyens, "La Conscience imaginée: Sur l'éliminativisme de Daniel Dennett," *Revue philosophique de Louvain* 98, no. 4 (2000): 773.

3. Jakob von Uexküll, *Streifzüge durch die Umwelten von Tieren und Menschen* (Berlin: Springer, 1934), 29.

4. Victor Petit, "Le Concept de milieu en amont et en aval de Simondon," in Jean-Hugues Barthélémy, *Cahiers Simondon*, no. 5 (Paris: L'Harmattan, 2013).

5. Uexküll, *Streifzüge durch die Umwelten*, 22.

6. Thomas Nagel, "What Is It Like to Be a Bat?" *Philosophical Review* 83 (1974): 435–450.

Supplement 3

1. Aristotle, *The Metaphysics*, trans. Hugh Treddenick (Cambridge, MA: Harvard University Press, 1975).

2. Gilles-Gaston Granger, *Le Probable, le possible et le virtuel: Essai sur le rôle du non-actuel dans la pensée objective* (Paris: Odile Jacob, 1995), 13.

3. Philippe Quéau, *Le Virtuel: Vertus et vertiges* (Seyssel: Champ Vallon, 1993), 26.

4. Pierre Lévy, *Qu'est-ce que le virtuel?* (Paris: La Découverte, 1998), 13.

5. Sylvain Missonnier and Hubert Lisandre, eds., *Le Virtuel: La présence de l'absent* (Paris: EDK, 2003).

6. Lévy, *Qu'est-ce que le virtuel?* 19.

7. Nathan Jurgenson, "When Atoms Meet Bits: Social Media, The Mobile Web and Augmented Revolution," *Future Internet*, no. 4 (2012), 85.

8. Jurgenson, "When Atoms Meet Bits," 84.

9. Jurgenson, 86.

10. Maude Bonenfant, "Les Mondes numériques ne sont pas 'virtuels,'" *Revue des sciences sociales* no. 45 (2011): 60–67.

11. "Digital dualism" must be understood in the sense of "dualism in the digital age."

12. Jurgenson, "When Atoms Meet Bits," 84.

13. Didier Anzieu, "Le Moment de l'apocalypse," *La Nef*, no. 31 (1967): 127–132.

14. Sophie de Mijolla-Mellor, *Le Plaisir de la pensée* (Paris: Presses universitaires de France, 1992).

15. Anzieu, "Le Moment de l'apocalypse," 130.

16. Jean Laplanche and Jean-Bertrand Pontalis, *The Language of Psychoanalysis*, trans. by Donald Nicholson-Smith (London: Hogarth Press, 1973).

17. Sherry Turkle, *The Second Self: Computers and the Human Spirit* (New York: Simon & Schuster, 1984); *SecondLife.com*.

18. Sherry Turkle, *Life on the Screen: Identity in the Age of the Internet* (New York: Simon & Schuster, 1995), 23.

19. Neal Stimler and Stéphane Vial, "Digital Monism: Our Mode of Being at the Nexus of Life, Digital Media and Art," Theorizing the Web 2014, New York, April 26, 2014.

20. Paul Miller, "I'm Still Here: Back Online after a Year without the Internet," *Verge*, May 1, 2013, http://goo.gl/2Qmx5.

Part II: A Short Treatise on Design

Chapter 1

1. Quoted by Marie-Haude Caraës, "Pour une recherche en design," *Azimuts* 33 (2009): 39.

2. Kenya Hara, *Designing Design* (Baden: Lars Müller Publishers, 2008), 19.

3. This statement certainly deserves to be nuanced. It remains true, however, if we consider that the many works of international design researchers, in addition to being published mostly in English and therefore not very visible in France, have not yet generated "a global theory of design" as exists in the arts or in the sciences. Bruce Brown, Richard Buchanan, Carl DiSalvo, Dennis Doordan, and Victor Margolin rightly point this out in *Design Issues* 29, no. 2 (2013): 1.

4. Caraës, "Pour une recherche en design," 39. It is, however, necessary to acknowledge one exception: the work of Alain Findeli, a pioneer in the epistemology of design, whose publications are poorly known in France because they are mostly in the English.

5. Caraës, "Pour une recherche en design," 40.

6. Jean-Louis Frechin, "Interfaces: Un Rôle pour le design," in *Le Design de nos existences*, ed. Bernard Stiegler (Paris: Mille et une nuits, 2008), 255.

7. Sophie de Mijolla-Mellor, *Le Plaisir de pensée* (Paris: Presses universitaires de France, 1992). See the definition on 498: "the capacity shown by certain topics to derive pleasure from their intellection and their discursive activity."

Chapter 2

1. Jean Baudrillard, *The Consumer Society: Myths and Structures*, trans. Chris Turner (Thousand Oaks, CA: Sage, 2009).

2. Bruno Remaury, "Les Usages culturels du mot design," in *Le Design: Essais sur des théories et des pratiques*, 2nd ed., ed. Brigitte Flamand (Paris: Institut français de la mode: Éditions du Regard, 2006), 108.

3. Remaury, "Les Usages culturels du mot design," 110.

4. Vilém Flusser, *Vom Stand der Dinge: Eine kleine Philosophie des Design* (Göttingen: Steidl Verlag, 1993), 9.

5. Alexandra Midal, *Design: Introduction à l'histoire d'une discipline* (Paris: Pocket, 2009), 33–34.

6. Karl Marx and Frederick Engels, preface to the 1888 English Edition of *Manifesto of the Communist Party*, trans. Samuel Moore in cooperation with Frederick Engels, 1888, 8, https://www.marxists.org/archive/marx/works/download/pdf/Manifesto.pdf.

7. Quoted in Nikolaus Pevsner, *The Sources of Modern Architecture and Design* (New York: Oxford University Press, 1968), 13.

8. William Morris, "How We Live, How We Might Live," accessed May 11, 2018, at https://www.marxists.org/archive/morris/works/1884/hwl/hwl.htm.

9. The name of a Parisian shop established in late 1895 (Pevsner, *Sources of Modern Architecture*, 43).

10. Hal Foster, *Design and Crime: And Other Diatribes* (London: Verso, 2002), 13.

11. Midal, *Design*.

12. Conference given on October 30, 1889, in Edinburgh, published under the title "Applied Arts Today," accessed May 14, 2018, at https://www.marxists.org/archive/morris/works/1889/today.htm.

13. Étienne Souriau, *Vocabulaire d'esthétique* (Paris: Presses universitaires de France, 1990), 146.

14. See Andrea Branzi, *Qu'est-ce que le design?* (Paris: Gründ, 2009), 126.

15. See, for example, Danielle Quarante, *Eléments de design industriel* (Paris: Polytechnica, 1994), 21: "We generally date the beginning of the history of design around the late eighteenth century with the appearance of the steam engine." Let's not confuse design and industry. But, there's worse, from the beautiful pen of Kenya Hara: "Design began at the very moment man began to use tools." Let's not confuse design and technique or technology. Kenya Hara, *Designing Design* (Baden: Lars Müller Publishers, 2008), 412.

16. Raymond Loewy, *Never Leave Well Enough Alone* (New York: Simon & Schuster, 1951), 236.

17. See Midal, *Design*, 169.

18. Souriau, *Vocabulaire d'esthétique*.

19. "Loi no. 94-665 du 4 août 1994 relative à l'emploi de la langue française," https://www.legifrance.gouv.fr/affichTexte.do?cidTexte=LEGITEXT000005616341.

Chapter 3

1. Alexandra Midal, *Design: Introduction à l'histoire d'une discipline* (Paris: Pocket, 2009), 143.

2. Raymond Loewy, *Never Leave Well Enough Alone* (New York: Simon & Schuster, 1951), 212.

3. Loewy, *Never Leave Well Enough Alone*, 237.

4. "Philippe Starck Meditates on Design," TED, Monterey, CA, March 2007, accessed May 14, 2018, at http://goo.gl/3b1514.

5. Loewy, *Never Leave Well Enough Alone*, 10.

6. Loewy, 210.

7. Loewy, 211.

8. John Maeda, *The Laws of Simplicity* (Cambridge, MA: MIT Press, 2006).

9. Bernard Stiegler, *Aimer, s'aimer, nous aimer: Du 11 septembre au 21 avril* (Paris: Galilée, 2003), 22.

10. Benoît Heilbrunn, "Le Marketing à l'épreuve du design," in *Le Design: Essais sur des théories et des pratiques*, ed. Brigitte Flamand (Paris: Éditions du Regard, 2013), 277–294.

11. Heilbrunn, "Le Marketing à l'épreuve du design."

12. Heilbrunn.

13. All of these quotations are taken from a Parisian global design agency's website chosen at random on the Web: accessed May 2010 and January 2014, http://www.pulp--design.com.

14. Association Design Council, accessed May 2010 at http://www.adc-asso.com.

15. Hal Foster, *Design and Crime: And Other Diatribes* (London: Verso, 2002), 18.

16. Foster, *Design and Crime*, 17.

17. Foster, 18.

18. Frédéric Beigbeder, *£9.99: A Novel*, trans. Adriana Hunter (London: Picador, 2002), 5–7.

19. Victor Papanek, *Design for the Real World* (Chicago: Academy Chicago Publishers, 1985), ix.

20. Papanek, *Design for the Real World*, xiii.

21. Papanek, i–ix.

22. Papanek, ix.

23. Vilém Flusser, *The Shape of Things: A Philosophy of Design*, trans. Anthony Mathews (London: Reaktion Books, 1999), 9.

24. Flusser, *Shape of Things*, 11.

25. Flusser, 27.

26. Flusser, 29.

27. Flusser, 29.

28. Flusser, 30.

29. Flusser, 31, 33.

Chapter 4

1. See Stéphane Vial, "Designers and Paradoxical Injunctions: How Designerly Ways of Thinking Are Faced with Contradiction," paper presented at the Fifth International Congress of International Association of Societies of Design Research, Shibaura Institute of Technology, Tokyo, August 26–30, 2013, http://design-cu.jp/iasdr2013/papers/2124-1b.pdf.

2. Harold F. Searles, "The Effort to Drive the Other Person Crazy: An Element in the Aetiology and Psychotherapy of Schizophrenia," *Psychology and Psychotherapy* 33, no. 1 (1959): 1–18.

3. Brigitte Flamand, ed., *Le Design: Essais sur des théories et des pratiques* (Paris: Éditions du Regard, 2006, 2013).

4. Alexandra Midal, *Design: Introduction à l'histoire d'une discipline* (Paris: Pocket, 2009), 138.

5. Ettore Sottsass, "Mi dicono che sono cattivo," *Scritti 1946–2001* (Vicenza: Neri Pozza, 2002), 242–245.

6. Quoted in Midal, *Design*, 144.

7. Midal, *Design*, 145.

8. Midal, 146.

9. Clémence Mergy, "Design, conditions d'apparition and légitimation," *Azimuts* 30 (2008): 91.

10. In the digital age, citizens' and users' growing participation in decision making, design, and production processes tends to minimize this aspect.

11. Quoted in Midal, *Design*, 145.

Chapter 5

1. Vilém Flusser, *The Shape of Things: A Philosophy of Design*, trans. Anthony Mathews (London: Reaktion Books, 1999), 11.

2. Brigitte Flamand, ed., *Le Design: Essais sur des théories et des pratiques* (Paris: Éditions du Regard, 2006).

3. Bruno Remaury, "Les Usages culturels du mot design," in *Le Design*, 101.

4. Kenya Hara, *Designing Design* (Baden: Lars Müller Publishers, 2008), 467.

5. Søren Kierkegaard, "Reduplication Is Being What One Says," in Perry D. LeFevre, *The Prayers of Kierkegaard* (Chicago: University of Chicago Press, 1956), 189.

6. Cf. Postscript, definitions 1 and 2.

7. Cf. Stéphane Vial, "Le Geste de design et son effet: Vers une philosophie du design," *Figures de l'art*, October 25, 2013, Presses Universitaires de Pau et des pays de l'Adour, 93–105.

8. Cf. Postscript, proposition II.

9. Cf. Postscript, axiom III.

10. Jean-Pierre Séris, *La Technique* (Paris: Presses universitaires de France, 2000), 267.

11. Quoted in Danielle Quarante, *Éléments de design industriel* (Paris: Polytechnica, 1994), 56.

12. Adolf Loos, *Ornament and Crime: Selected Essays* trans. Michael Mitchell (Riverside, CA: Ariadne Press, ca. 1998).

13. Hara, *Designing Design*, 243.

14. Loos, *Ornament and Crime*, 78.

15. Bernard Stiegler, "Du design comme sculpture sociale," in *Le Design*, 351.

16. Alain Findeli, "Searching for Design Research Questions: Some Conceptual Clarifications," in *Questions, Hypotheses and Conjectures: Discussions on Projects by Early Stage and Senior Design Researchers* (Bloomington, IN: IUniverse, 2010), 292.

Chapter 6

1. Sigmund Freud, "Creative Writing and Day-Dreaming" (1908), http://www.kleal.com/AP12%20member%20area%20pd2%202013/Freud%20and%20Frye.pdf.

2. See the exhibit "Our Body: À corps ouvert," Paris, February 12–May 10, 2009.

3. See http://theconversation.com/paola-antonelli-interview-design-has-been-misconstrued-as-decoration-21148.

4. Clémence Mergy, "Design, conditions d'apparition et légitimation,"*Azimuts* 30 (2008): 92.

5. Mergy, "Design, conditions," 92.

6. Video interview with Patrick Jouin, May 5, 2010, available at the Center Pompidou website: http://www.centrepompidou.fr/.

7. Kenya Hara, *Designing Design* (Baden: Lars Müller Publishers, 2008), 24.

8. Herbert A. Simon, *The Sciences of the Artificial* (Cambridge, MA: MIT Press, 1981), 129.

9. Alain Findeli, "Le Design, discipline scientifique? Une esquisse programmatique," presented at the symposium Les Ateliers de la recherche en design, University of Nîmes, November 13–14, 2006.

10. See also Stéphane Vial, "Design and Creation: Outline of a Philosophy of Modelling," *Wikicreation*, June 2016, http://wikicreation.fr/en/articles/270.

Chapter 7

1. Tim Brown, "Design Thinking," *Harvard Business Review*.June 2008): 84ff.

2. I am relying here on two lectures delivered by Tim Brown: one at MIT in March 2006, "Innovation through Design Thinking" (http://goo.gl/eBmaBm); the other a TED Global2009 talk, "Tim Brown Urges Designers to Think Big" (http://goo.gl/rUdeA).

3. Alain, *Système des Beaux-Arts* (Paris: Gallimard, 1983), I, 7.

4. Don Norman, "Design Thinking: A Useful Myth," *Core77*, June 25, 2010, http://goo.gl/veL3.

5. Bill Moggridge, "Design Thinking: Dear Don ...," *Core77*, August 2, 2010, http://goo.gl/ay2I.

6. Don Norman, "Rethinking Design Thinking," *Core77*, March 19, 2013, http://goo.gl/DUu2va.

Chapter 8

1. John Maeda, *The Laws of Simplicity* (Cambridge, MA: MIT Press, 2016), 1.

2. Cf. Stéphane Vial, *Being and the Screen: How the Digital Changes Perception*, this volume.

3. Cf. Stéphane Vial, "Qu'appelle-t-on 'design numérique?'" *Interfaces numériques* 1, no. 1 (2012): 91–106.

4. Sherry Turkle, *Life on the Screen: Identity in the Age of the Internet* (New York: Simon & Schuster, 1995).

5. Quoted in Sherry Turkle, *Simulation and Its Discontents* (Cambridge, MA: MIT Press, 2009), 88.

6. Turkle, *Simulation and Its Discontents*, 43–44.

7. Turkle, 44.

8. Turkle, 46.

9. Pierre von Meiss, *De la cave au toit: Témoignage d'un enseignement d'architecture* (Lausanne: Polytechnic Press and University Romandes, 1991), 71.

10. Jean-Jacques Rousseau, *Émile*. [My translation.—Trans.]

11. Kenya Hara, *Designing Design* (Baden: Lars Müller Publishers, 2008), 429–430.

12. Hara, *Designing Design*.

13. Since the first French-language edition of this book in 2010, 3D printers have become considerably more available.

14. Bill Moggridge, *Designing Interactions* (Cambridge, MA: MIT Press, 2007). See also the online video at http://www.designinginteractions.com/chapters/introduction.

15. Giuseppe Fioretti and Giancarlo Carbone, "Integrate Business Modeling and Interaction Design," June 2007, accessed May 2010 and January 2014 at http://www.ibm.com/developerworks/library/ws-soa-busmodeling/index.html.

16. Moggridge, *Designing Interactions*.

17. Proposed by Xerox Corporation in 1981, three years before the first Apple Macintosh, the Xerox Star was the first computer to offer a graphical user interface based on WIMP (windows, icons, menus, mouse) technology.

18. Cf. Bill Verplank, accessed May 2010 and January 2014 at http://www.billverplank.com/professional.html.

19. Interviewed on June 2, 2009, at the celebration of the Atelier de design numérique's tenth anniversary at Ensci, accessed May 2010: http://www.ensci.com/recherche-et-editions/editions/paroles/interviews/article/944/.

20. Jean-Louis Frechin, "Interfaces: Un Role pour le design" in *Le Design de nos existences*, ed. Bernard Stiegler (Paris: Mille et une nuits, 2008), 257.

21. Richard M. Stallman and Paul Mathias, "Informatiques et liberté," *Rue Descartes*, no. 55 (2007): 72–83, https://www-cairn-info.ezp-prod1.hul.harvard.edu/revue-rue-descartes-2007-1-page-72.htm.

22. Franck Varenne, *Qu'est-ce que l'informatique?* (Paris: Vrin, 2009).

23. Hara, *Designing Design*, 431.

24. Interaction Design Association, accessed May 2010 and January 2014 at http://www.ixda.org/about/ixda-mission.

25. Designers Interactifs, *Petit dictionnaire du design numérique* (Paris: Designers Interactifs, 2009), 15.

26. Designers Interactifs, *Petit dictionnaire*, 18.

Postscript

1. Video interview available at the Rue89 website: accessed January 2014 at http://www.rue89.com/artnet/2010/05/01/patrick-jouin-grand-chef-nantais-du-designparisien-149784.

Bibliography

Alain. *Système des Beaux-Arts*. Paris: Gallimard, 1983.

Aristotle. *The Metaphysics*. Translated by Hugh Treddenick. Cambridge, MA: Harvard University Press, 1975.

Aristotle. *Nicomachean Ethics*. Translated by H. Rackham. Cambridge, MA: Harvard University Press, 1934.

Bachelard, Gaston. *L'Activité rationaliste de la physique contemporaine*. Paris: Presses universitaires de France, 1965.

Bachelard, Gaston. *Atomistic Intuitions: An Essay on Classification*. Translated by Roch C. Smith. Albany: SUNY Press, 2018.

Bachelard, Gaston. *Études*. Paris: Vrin, 2002.

Bachelard, Gaston. *The Formation of the Scientific Mind*. Translated and annotated by Mary McAllester Jones. Manchester: Clinamen, 2002.

Bachelard, Gaston. *The New Scientific Spirit*. Translated by Arthur Goldhammer. Boston: Beacon Press, 1984.

Bachelard, Gaston. "Noumène et microphysique." *Recherches philosophiques* 1 (1931–1932): 55–65.

Bachelard, Gaston. *The Psychoanalysis of Fire*. Translated by Alan C. M. Ross. Boston: Beacon Press, 1964.

Baudrillard, Jean. *The Consumer Society: Myths and Structures*. Translated by Chris Turner. Thousand Oaks, CA: Sage, 1998.

Beahm, George. *I, Steve: Steve Jobs in His Own Words*. Chicago: B2 Books, 2011.

Beaune, Jean-Claude. *L'Automate et ses mobiles*. Paris: Flammarion, 1980.

Beaune, Jean-Claude. "Philosophy of Technology in France in the Twentieth Century: Overview and Current Bibliography." *Research in Philosophy and Technology* 2 (1979): 273–292.

Beaune, Jean-Claude. *La Technologie*. Paris: Presses universitaires de France, 1972.

Beigbeder, Frédéric. *£9.99: A Novel*. Translated by Adriana Hunter. London: Picador, 2002.

Benjamin, Walter. "A Short History of Photography." 1931. Translated by Stanley Mitchell. *Screen*, March 1, 1972, 5–26, https://doi.org/10.1093/screen/13.1.5.

Benjamin, Walter. "The Work of Art in the Age of Its Technological Reproducibility." Translated by Michael W. Jennings. *Grey Room*, no. 39 (2010): 11–37, https://www.mitpressjournals.org/doi/abs/10.1162/grey.2010.1.39.

Bergson, Henri. *Time and Free Will: Essay on the Immediate Data of Consciousness*. 1889. Translated by F. L. Pogson. London: George Allen & Unwin, 1921.

Besnier, Jean-Michel. *Les Théories de la connaissance*. Paris: Presses universitaires de France, 2011.

Beyaert-Geslin, A. "Formes de table, formes de vie. Réflexions sémiotiques pour vivre ensemble." *MEI: Médiation Et Information* 30–31 (2009): 99–110.

Bonenfant, Maude. "Les Mondes numériques ne sont pas 'virtuels.'" *Revue des sciences sociales*, no. 45 (2011): 60–67.

Branzi, Andrea. *Qu'est-ce que le design?* Paris: Gründ, 2009.

Breton, Philippe. *The Culture of the Internet and the Internet as Cult: Social Fears and Religious Fantasies*. Translated by David Bade. Duluth, MN: Litwin Books, 2011.

Breton, Philippe. *Une Histoire de l'informatique*. Paris: La Découverte, 1987.

Brown, Tim. "Design Thinking." *Harvard Business Review* 86, no. 6 (June 2008): 84–92.

Canguilhem, Georges. "Machine and Organism." In *The Knowledge of Life*. Translated by Stefanos Geroulanos and Daniela Ginsburg. New York: Fordham University Press, 2008.

Canguilhem, Georges. *The Normal and the Pathological*. Translated by Carolyn R. Fawcett. New York: Zone Books, 1991.

Caraës, Marie-Haude. "Pour une recherche en design." *Azimuts* 33 (2009).

Casilli, A. A. *Les Liaisons numériques: Vers une nouvelle sociabilité?* Paris: Le Seuil, 2010.

Casson, Herbert N. *The History of the Telephone*. Chicago: McClurg, 1910. http://etext.lib.virginia.edu/toc/modeng/public/CasTele.html.

Cauquelin, Anne. *L'Invention du paysage*. Paris: Presses universitaires de France, 2000.

Chabot, Pascal, and Gilbert Hottois, eds. *Les Philosophes et la Technique*. Paris: Vrin, 2003.

Cringeley, Robert X. *The Triumph of the Nerds: The Rise of Accidental Empires*. Directed by Paul Sen. TV documentary. Oregon Public Broadcasting, 1996.

Dagognet, François. *Éloge de l'objet*. Paris: Vrin, 1989.

Darras, Bernard. "Aesthetics and Semiotics of Digital Design: The Case of Web Interface Design." In *Proceedings of the INDAF International Conference*, 10–16. Incheon, Korea, 2009.

Darras, Bernard. "Ambition et création artistique assistée par ordinateur." In *Faire/voir et savoir: Connaissance de l'image, image et connaissance*, edited by Brigitte Poirier and Josette Sultan, 89–94. Paris: INRP, 1992.

Darras, Bernard. "Machines, complexité et ambition." In *Dessine-moi un pixel: Informatique et arts plastiques*, edited by J. Sultan and B. Tissor, 99–107. Paris: INRP/Centre Georges-Pompidou, 1991.

Darras Bernard, and Sarah Belkhamsa. "Les Objets communiquent-ils?" *MEI: Médiation Et Information* 30–31 (2009): 7–9.

de Broglie, Pauline, comtesse de Pange. *Comment j'ai vu 1900*. Paris: Grasset, 1962–1968.

Deleuze, Gilles, and Félix Guattari. *A Thousand Plateaus*. Translated by Brian Massumi. Minneapolis: University of Minnesota Press, 1987.

Eliade, Mircea. *The Sacred and the Profane. The Nature of Religion*. Translated by William R. Trask. New York: Harper & Row, 1961.

Ellul, Jacques. *The Technological System*. Translated by Joachim Neugroschel. New York: Continuum, 1980.

Findeli, A. "Searching for Design Research Questions: Some Conceptual Clarifications." In *Questions, Hypotheses, and Conjectures: Discussions on Projects by Early Stage and Senior Designers*, edited by Alain Findeli, Keith Russell, Rosan Chow, Wolfgang Jonas, and Geche Joost. Bloomington, IN: iUniverse, 2010.

Flamand, Brigitte, ed. *Le Design: Essais sur des théories et des pratiques*. Paris: Éditions du Regard, 2013.

Flusser, Vilém. *The Shape of Things: A Philosophy of Design*. Translated by Anthony Mathews. London: Reaktion Books, 1999.

Foster, Hal. *Design and Crime and Other Diatribes*. London: Verso, 2002.

Francastel, Pierre. *Art and Technology in the Nineteenth and Twentieth Centuries*. Translated by Randall Cherry. New York: Zone Books, 2000.

Frechin, Jean-Louis. "Interfaces: Un rôle pour le design." In *Le Design de nos existences*, edited by Bernard Stiegler. Paris: Mille et une nuits, 2008.

Freud, Sigmund. *On Creativity and the Unconscious: Papers on the psychology of Art, Religion, Love, Religion.* Selected, with introduction and annotations by Benjamin Nelson. New York: Harper Perennial Modern Classics, 2009.

Genvo, Sébastien. "Penser les phénomènes de 'ludicisation' du numérique: Pour une théorie de la jouabilité." *Revue des sciences sociales* 45 (2011): 68–77.

Gibson, William. *Neuromancer.* New York: Ace Books, 1984.

Giedion, Siegfried. *Mechanization Takes Command: A Contribution to Anonymous History.* Oxford: Oxford University Press, 1948.

Gille, Bertrand. *Engineers of the Renaissance.* Cambridge, MA: MIT Press, 1966.

Gille, Bertrand. *The History of Techniques.* Translated by P. Southgate and T. Williamson. Montreux: Gordon and Breach Scientific Publishers, 1986.

Granger, Gilles-Gaston. *Le Probable, le Possible et le Virtuel: Essai sur le rôle du non-actuel dans la pensée objective.* Paris: Odile Jacob, 1995.

Guillaud, Hubert. "Où va l'économie numérique? (1/3) Vers une innovation sans emplois?" InternetActu.net, February 1, 2012. http://www.internetactu.net/2012/02/01/ou-va-leconomie-numerique-13-vers-uneinnovation-sans-emplois.

Habermas, Jürgen. "Technology and Reification: Technology and Science as 'Ideology.'" Translated by Robin Celikates and Rahel Jaeggi (1968). In *The Habermas Handbook*, edited by Hauke Brunkhorst, Regina Kreide, and Cristina Lafont. New York: Columbia University Press, 2018.

Hadot, Pierre. *Philosophy as a Way of Life: Spiritual Exercises from Socrates to Foucault.* Oxford: Blackwell, 1995.

Hara, Kenya. *Designing Design.* Baden: Lars Müller Publishers, 2007.

Hegel, G. W. F. *Hegel's Aesthetics.* Edited and translated by T. M. Knox. Oxford: Oxford University Press, 2015.

Heidegger, Martin. "The Question Concerning Technology." Translated by William Lovitt. In *The Question Concerning Technology.* New York: Harper & Row, 1977.

Hottois, Gilbert. "Les Philosophes et la technique, les philosophes de la technique." In *Les Philosophes et la technique*, edited by Pascal Chabot and Gilbert Hottois, 13–23. Paris: Vrin, 2003.

Huyghe, Pierre-Damien. *Du commun: Philosophie pour la peinture et le cinéma.* Belval: Circé, 2002.

Huyghe, Pierre-Damien, ed. *L'Art au temps des appareils.* Paris: L'Harmattan, 2005.

Huyghe, Pierre-Damien. "Le Devenir authentique des techniques." Conférence au Centre national de la recherche technologique. Rennes, 2004. http://pierredamienhuyghe.fr/documents/textes/huyghethomson.pdf.

Huyghe, Pierre-Damien. *Le Différend esthétique*. Belval: Circé, 2004.

Huyghe, Pierre-Damien. "Introduction au dossier 'Temps et appareils.'" *Plastik*, no. 3 (Fall 2003): 4–6.

Huyghe, Pierre-Damien. *Modernes sans modernité*. Paris: Éditions Lignes, 2009.

Ibnelkaïd, Samira. "Identité et altérité par écran: Modalités de l'intersubjectivité en interaction numérique." PhD diss., Université Lumière, 2016. https:/transphanie.com.

Jones, Sheilla. *The Quantum Ten: A Story of Passion, Tragedy, Ambition, and Science*. Oxford: Oxford University Press, 2008.

Jouin, Patrick. "La Substance du design." Video interview, Centre Georges-Pompidou, Paris, February 17–May 24, 2010. http://www.centrepompidou.fr/presse/video/20100119-jouin/.

Jurgenson, Nathan. "Fear of Screens." *New Inquiry*, January 25, 2016. https://thenewinquiry.com/fear-of-screens/.

Jurgenson, Nathan. "The IRL Fetish." *New Inquiry*, June 28, 2012. https://thenewinquiry.com/the-irl-fetish/.

Jurgenson, Nathan. "When Atoms Meet Bits: Social Media, the Mobile Web and Augmented Revolution." *Future Internet* 4 (2012).

Kant, Immanuel. *Critique of Judgment*. Translated by J. H. Bernard. London: Macmillan, 1914. http://lfoll.s3.amazonaws.com/titles/1217/Kant_0318_EBk_v6.0.pdf.

Krainin, Julian, and Michael R. Lawrence. *Memory and Imagination: New Pathways to the Library of Congress*. Baltimore: Michael Lawrence Films and Krainin Productions, 1990.

Kuhn, Thomas S. *The Structure of Scientific Revolutions*. Chicago: University of Chicago Press, 1962.

Lalande, André. *Vocabulaire technique et critique de la philosophie*. Paris: Presses universitaires de France, 1960.

Latour, Bruno. *Aramis, or The Love of Technology*. Translated by Catherine Porter. Cambridge, MA: Harvard University Press, 1996.

Latour, Bruno. *We Have Never Been Modern*. Translated by Catherine Porter. Cambridge, MA: Harvard University Press, 1993.

Latour, Bruno. "A Cautious Prometheus? A Few Steps toward a Philosophy of Design (with Special Attention to Peter Sloterdijk)." In *Proceedings of the 2008 Annual International Conference of the Design History Society*, edited by F. Hackne, J. Glynne, and V. Minto, 2–10. Universal Publishers, 2009. E-book. http://www.brunolatour.fr/sites/default/files/112-DESIGNCORNWALLGB.pdf.

Lavoisier, Antoine-Laurent. *Elements of Chemistry, in a New Systematic Order, Containing all the Modern Discoveries.* Translated by Robert Kerr. Edinburgh: Printed for William Creech, 1740. http://www.gutenberg.org/cache/epub/30775/pg30775.txt.

Leleu-Merviel, Sylvie. "Les désarrois des 'Maîtres du sens' à l'ère du numérique." In *Hypertextes, hypermédias: Créer du sens à l'ère numérique, H2PTM'03*, 17–34. Lavoisier, 2003. http://halshs.archives-ouvertes.fr/hal-00467743.

Leleu-Merviel, Sylvie, and Philippe Useille. "Quelques revisions du concept d'information." In *Problématiques émergentes dans les sciences de l'information*, edited by Fabrice Papy, 25–56. Lavoisier, 2008. http://hal.archives-ouvertes.fr/hal-00695777.

Le Moigne, Jean-Louis. *Les Épistémologies constructivistes.* Paris: Presses universitaires de France, 2007.

Leroux, Yann. "En lisant Stéphane Vial." *Mondes numériques*, May 8, 2009. Google Groups.

Leroux, Yann. "Psychodynamique des groupes sur le réseau Internet." PhD diss., Université Paris X Nanterre, 2010.

Lessig, Lawrence. *Free Software, Free Society: Selected Essays of Richard M. Stallman.* 2002. http://www.gnu.org/philosophy/lessig-fsfs-intro.fr.html.

Lévi-Strauss, Claude. *Entretiens avec Georges Charbonnier.* Paris: Union générale d'éditions, 1998.

Lévy, Pierre. *Cyberculture.* Translated by Robert Bononno. Minneapolis: University of Minnesota Press, 2001.

Lévy, Pierre. *De la programmation comme un des beaux-arts.* Paris: La Découverte, 1992.

Lévy, Pierre. *La Machine univers: Création, cognition et culture informatique.* Paris: La Découverte, 1987.

Lévy, Pierre. "Le Médium algorithmique." *Sociétés* 3, no. 129 (2015): 79–96.

Lévy, Pierre. *Qu'est-ce que le virtuel?* Paris: La Découverte, 1998.

Lévy, Pierre. *The Semantic Sphere.* Hoboken, NJ: Wiley, 2011.

Levy, Steven. *The Perfect Thing: How the iPod Shuffles Trade, Culture and Coolness.* New York: Simon & Schuster, 2006.

Lœwy, Raymond. *Never Leave Well Enough Alone.* New York: Simon & Schuster, 1951.

Lombard, Geneviève. "Le non-virtualisable de la psychanalyse." Inconscient.net, September 26, 2007. http://inconscient.net/non_virtualisable.htm.

Loos, Adolf. *Ornament and Crime: Selected Essays.* Translated by Michael Mitchell. Riverside, CA: Ariadne Press, 1998.

Maeda, John. *The Laws of Simplicity*. Cambridge, MA: MIT Press, 2016.

Manovich, Lev. *Software Takes Command*. New York: Bloomsbury, 2013.

Marcuse, Herbert. *One-Dimensional Man: Studies in the Ideology of Advanced Industrial Society*. Boston: Beacon Press, 1964.

Mathias, Paul. *Qu'est-ce que l'Internet?* Paris: Vrin, 2009.

Meiss, Pierre von. *De la cave au toit: Témoignage d'un enseignement d'architecture*. Lausanne: Presses polytechniques et universitaires romandes, 1991.

Mergy, Clémence. "Design, conditions d'apparition et légitimation." *Azimuts* 30 (2008).

Midal, Alexandra. *Design: Introduction à l'histoire d'une discipline*. Paris: Pocket, 2009.

Mijolla-Mellor, Sophie de. *Le Plaisir de pensée*. Paris: Presses universitaires de France, 1992.

Milon, Alain. *La Réalité virtuelle: Avec ou sans le corps?* Paris: Autrement, 2005.

Moggridge, Bill. "Design Thinking: Dear Don ..." *Core77*, August 2, 2010. https://www.core77.com/posts/17042/design-thinking-dear-don-17042.

Moggridge, Bill. *Designing Interactions*. Cambridge, MA: MIT Press, 2007.

Morris, William. "How We Live, How We Might Live." Accessed May 11, 2018, at https://www.marxists.org/archive/morris/works/1884/hwl/hwl.htm.

Morozov, Evgeny. *To Save Everything, Click Here: The Folly of Technological Solutionism*. New York: Public Affairs Press, 2013.

Mumford, Lewis. *Technics and Civilization*. New York: Harcourt & Brace, 1934.

Musso, Pierre. *Critique des réseaux*. Paris: Presses universitaires de France, 2003.

Musso, Pierre, Laurent Ponthou, and Éric Seulliet. *Fabriquer le futur, 2: L'Imaginaire au service de l'innovation*. Paris: Pearson Education France, 2007.

Norman, Don. "Design Thinking: A Useful Myth." *Core77*, June 25, 2010. https://www.core77.com/posts/16790/design-thinking-a-useful-myth-16790.

Norman, Don. "Rethinking Design Thinking." *Core77*, March 19, 2013. https://www.core77.com/posts/24579/rethinking-design-thinking-24579#more.

Nietzsche, Friedrich. "On the Uses and Disadvantages of History for Life." Translated by Ian Johnstone. http://www.leudar.com/library/On%20the%20Use%20and%20Abuse%20of%20History.pdf.

Nietzsche, Friedrich. *Twilight of the Idols*. In *The Portable Nietzsche*, edited and translated by Walter Kaufmann, 463–563. New York: Penguin Books, 1976.

Nova, Nicolas. "Famous User Figures in the History of HCI." February 18, 2010. http://www.nicolasnova.net/pasta-and-vinegar/2010/02/18/famous-user-figures-in-the-history-of-hci.

Papanek, Victor. *Design for the Real World: Human Ecology and Social Change*. Chicago: Academy Chicago, 1985.

Parrochia, Daniel. "L'Internet et ses représentations." In *Philosophies entoilées*, 10–20. Rue Descartes 55. Paris: Presses universitaires de France, 2007.

Parrochia, Daniel. *Penser les réseaux*. Seyssel: Champ Vallon, 2001.

Pevsner, Nikolaus. *The Sources of Modern Architecture Design*. New York: Oxford University Press, 1968.

Proulx, Serge, and Guillaume Latzko-Toth. "La Virtualité comme catégorie pour penser le social: L'usage de la notion de communauté virtuelle." *Sociologie et Société* 32, no. 2 (2000): 99–122.

Quarante, Danielle. *Éléments de design industriel*. Paris: Polytechnica, 1994.

Quéau, Philippe. *Éloge de la simulation: De la vie des langages à la synthèse des images*. Seyssel: Champ Vallon, "Milieux," 1986.

Quéau, Philippe. *Metaxu: Théorie de l'art intermédiaire*. Seyssel: Champ Vallon/INA, 1989.

Quéau, Philippe. "La Pensée virtuelle." *Réseaux* 61 (September–October 1993): 67–78.

Quéau, Philippe. *Le Virtuel: Vertus et vertiges*. Seyssel: Champ Vallon, "Milieux," 1993.

Remaury, Bruno. "Les Usages culturels du mot design." In *Le Design: Essais sur des théories et des pratiques*, edited by Brigitte Flamand. Paris: Institut français de la mode, Éditions du Regard, 2006.

Rifkin, Jeremy. *The Third Industrial Revolution: How Lateral Power Is Transforming Energy, the Economy, and the World*. New York: Palgrave Macmillan, 2011.

Roger, Alain. *Court traité du paysage*. Paris: Gallimard, 1997.

Searles, Harold F. "The Effort to Drive the Other Person Crazy: An Element in the Aetiology and Psychotherapy of Schizophrenia." *Psychology and Psychotherapy* 33, no. 1 (1959): 1–18.

Séris, Jean-Pierre. *La Technique*. Paris: Presses universitaires de France, 2000.

Serres, Michel. "Ce n'est pas une crise, c'est un changement de monde." *Journal du dimanche*, December 30, 2012. https://www.lejdd.fr/Economie/Serres-Ce-n-est-pas-une-crise-c-est-un-changement-de-monde-583645-3134546.

Serres, Michel. *Thumbelina: The Culture and Technology of Millennials*. Translated by Daniel W. Smith. London: Rowman & Littlefield, 2014.

Simon, Herbert A. *The Sciences of the Artificial*. Cambridge, MA: MIT Press, 1981.

Simondon, Gilbert. *On the Mode of Existence of Technical Objects*. Translated by Cécile Malaspina and John Rogove. Minneapolis: University of Minnesota Press, 2017.

Simondon, Gilbert. "Psycho-sociologie de la technicité." *Bulletin de l'École pratique de psychologie et de pédagogie* 2 (1960): 127–140.

Sloterdijk, Peter. "The Domestication of Being." In *Not Saved: Essays after Heidegger*. Translated by Ian Alexander Moore and Christopher Turner. Cambridge: Polity Press, 2017.

Sloterdijk, Peter. "Foreword to the Theory of Spheres." In *Cosmograms*, edited by Melik Ohanian and Jean-Christophe Royoux, 223–240. New York: Lukas & Sternberg, 2005.

Sottsass, Ettore, Milco Carboni, and Barbara Radice. *Scritti 1946–2001*. Vicenza: Neri Pozza, 2002.

Souriau, Étienne. *Vocabulaire d'esthétique*. Paris: Presses universitaires de France, 1990.

Stallman, Richard M., and Paul Mathias. "Informatiques et liberté." *Philosophies entoilées* 55 (2007): 72–83.

Stiegler, Bernard. *Aimer, s'aimer, nous aimer*. Paris: Galilée, 2003.

Stiegler, Bernard. "Du design comme sculpture sociale." In *Le Design: Essais sur des théories et des pratiques*, edited by Brigitte Flamand. Paris: Éditions du Regard, 2006.

Stiegler, Bernard, ed. *Le Design de nos existences*. Paris: Mille et une nuits, 2008.

Stiegler, Bernard. *De la misère symbolique*. Vol. 1: *L'Époque hyper-industrielle*. Paris: Galilée, 2004.

Sullivan, Louis H. "The Tall Office Building Artistically Considered." *Lippincott's Magazine* 57 (March 1896). http://academics.triton.edu/faculty/fheitzman/tallofficebuilding.html.

Thackara, John. *In the Bubble: Designing in a Complex World*. Cambridge, MA: MIT Press, 2005.

Tisseron, Serge. *Rêver, fantasmer, virtualiser: Du virtuel psychique au virtuel numérique*. Paris: Dunod, 2012.

Tisseron, Serge. *Virtuel, mon amour: Penser, aimer, souffrir, à l'ère des nouvelles technologies*. Paris: Albin Michel, 2008.

Triclot, Mathieu. *Philosophie des jeux vidéo*. Paris: La Découverte, 2011.

Turkle, Sherry. *Alone Together: Why We Expect More from Technology and Less from Each Other*. New York: Basic Books, 2012.

Turkle, Sherry. *Life on the Screen: Identity in the Age of the Internet.* New York: Simon & Schuster, 1995.

Turkle, Sherry. *Simulation and Its Discontents.* Cambridge, MA: MIT Press, 2009.

Varenne, Franck. *Qu'est-ce que l'informatique?* Paris: Vrin, 2009.

Varenne, Franck, and Marc Silberstein. *Modéliser et simuler: Épistémologies et pratiques de la modélisation et de la simulation.* Paris: Éditions Matériologiques, 2013.

Vermaas, Pieter, and Stéphane Vial, eds. *Advancements in the Philosophy of Design.* Dordrecht: Springer, 2018.

Vial, Stéphane. "Contre le virtuel: Une déconstruction." *MEI: Médiation Et Information* 37 (2013).

Vial, Stéphane. "Designers and Paradoxical Injunctions: How Designerly Ways of Thinking Are Faced with Contradiction." Paper presented at the Fifth International Congress of International Association of Societies of Design Research, Shibaura Institute of Technology, Tokyo, August 26–30, 2013. http://design-cu.jp/iasdr2013/papers/2124-1b.pdf.

Vial, Stéphane. "Il était une fois 'pp7,' ou la naissance d'un groupe sur l'Internet: Retour sur la socialisation en ligne d'une communauté étudiante." *Réseaux* 6, no. 164 (2010): 51–70.

Vial, Stéphane. "Le Geste de design et son effet: Vers une philosophie du design." *Figures de l'art: Revue d'études esthétiques,* no. 26 (2013).

Vial, Stéphane. "Qu'appelle-t-on 'design numérique'?" *Interfaces numériques* 1, no. 1 (2012): 91–106.

Vial, Stéphane. "There Is No Difference between the 'Real' and the 'Virtual': A Brief Phenomenology of Digital Revolution." Speech at Theorizing the Web 2013, Graduate Center, City University of New York, March 2, 2013. https://fr.slideshare.net/reduplikation/vial-ttw13reloadedhigh.

Vignola, Robert. *Allô! La merveilleuse aventure du téléphone.* Paris: CZ Créations, 2000.

Volle, Michel. "Comprendre la croissance à l'heure de l'informatisation de la société." In *Une croissance intelligent,* edited by Philippe Lemoine. Paris: Descartes & Cie, 2012.

Volle, Michel. *Économie des nouvelles technologies.* Paris: Economica, 1999.

Volle, Michel. *De l'informatique: Savoir vivre avec l'automate.* Paris: Economica, 2006.

Wilde, Oscar. "The Decay of Lying." In his *Intentions.* New York: Brentano's, 1905.

Index

Actor-network theory, 126, 150
Actual, the, 83, 84
Actuality (*energeia*), 67, 83, 84
  defined, 67, 83
Actualization, 67, 71
  "reverse," 142
Advertising, 164
Advertising design, 167
Aesthetic constructivism, 51. *See also* Constructivism
Aesthetic decline, 54
"Aesthetic logic," 57
Aesthetics, 161, 170. *See also* Industrial aesthetics
  Pierre-Damien Huyghe and, 56–59, 211n43, 211n47
  photography and, 54–55
Aesthetics (Hegel), 109
Aesthetic values, 56
Allegory of the cave, Plato's, 82, 145
Allen, Woody, 101
Anti-Design, 171
Antonelli, Paola, 180
Apelles, 96
Apparatus
  concept of, 56, 57, 59
  defined, 57
  dialectic of, 59 (*see also* Dialectics of the apparatus and of appearance)
Apparition, uniqueness of, 127

Appearance
  dialectic of, 59 (*see also* Dialectics of the apparatus and of appearance)
  technology and, 58–59
Apple, 20, 21, 34, 175. *See also* Jobs, Steve
Apple II, 30, 89
Applied arts, 153, 159, 186, 192
Applied design, 199
Arab Spring, 143
Architecture (and architects), 113, 158–161, 170, 188, 190
  design and, 188–191, 199
  digital revolution and, 187, 188
  modern, 176
Aristotle, 66–67, 135, 142
Art, 51–52. *See also* Aesthetics; Benjamin, Walter: art, technology, and
  vs. design, 181
Artificial intelligence, 87–88
Art nouveau, 158–160, 166
"Aura," 54, 64, 82. *See also* Phenomenological aura
  Walter Benjamin and the, 54–56, 83, 127, 129
  decline of the, 54–56
  restoring the aura of things, 55
Automated technological system, 36. *See also* Contemporary technological system

Automation, 29, 33–36
Avatar, 134

Bachelard, Gaston, 7, 47, 48, 75–78
  epistemological constructivism, 44, 47
  The New Scientific Spirit, 44, 45, 53
  "Noumenon and Microphysics," 44, 53
  phenomeno-technology and, 14, 44–46, 53, 55, 58
  on quantum physics, 44–46, 82–84
  on science and philosophy, 8, 14
Baudrillard, Jean, 15, 156
Bauhaus, 160–161, 164
Beaune, Jean-Claude, 81, 123, 208n40
Beauty, 176. *See also* Aesthetics
Beauty effect, 175
Behrens, Peter, 19, 160, 161
Beigbeder, Frédéric, 166
Being, 9. *See also* Existence
Being-in-the-world, 9, 52–53, 62, 65, 80, 121
  ontophany and, 50–51, 63, 112, 122, 138
  technology and, 52–53, 63, 125, 138
Being-in-the-world-with-objects, 126
Being-there (*Dasein*), 9, 52–53, 138, 142. *See also* Being-in-the-world
Bell, Alexander Graham, 53, 60
Benjamin, Walter, 56, 60
  art, technology, and, 53–58
  aura and, 54–56, 83, 127, 129
  on perception, 211n42
  phenomeno-technology and, 53, 55, 56, 58, 59
  photography and, 53–57, 129
  writings, 53
Bergson, Henri, 21
Berners-Lee, Tim, 33, 88, 89
Beyaert-Geslin, Anne, 114, 116
Big data, 4, 7, 31
Big systems (IBM), 34

Bonenfant, Maude, 143
Bouroullec, Ronan, 116, 219n19
Branzi, Andrea, 170
Brown, Tim, 183–185
Bugs, software, 94–96

Calculation, 33–34
  culture of, 73
Callimorphic effect, xix, 150, 175–176, 178. *See also* Design effect(s): types and dimensions of
Callimorphism, 176
"Callocentrism," 153, 161
Cameras, 55–57. *See also* Photography
Canguilhem, Georges, 17, 18
Caraës, Marie-Haude, 153
Casilli, Antonio A., 97–99, 107
Casson, Herbert N., 62
CATIA software, 190
Cauquelin, Anne, 51
Cinematographic images, 92
Cogito, ergo sum ("I think, therefore I am"), 183
Cognitive virtual, 83
Cole, Henry, 19, 20, 157, 159, 160
Communications design, 165, 166
  defined, 165
Compiler, 87
Computation, 30
Computational materials, 117
Computational matter, 86, 102–105
  bugs and, 95, 96
  design effects and, 118
  digital design and, 117, 118
  digital interfaces and, 120
  digital revolution and, 85
  fluidity and, 105
  human artifacts and, 120
  humans and, 91, 92
  nature of, 86, 89, 92, 96, 105, 117
  as noumenal, 86, 91, 94
  ontophanic properties and capabilities of, 118, 120

Index

and the virtual, 94, 102
working with, 88
Computational simulation, 69, 73, 74
"Computer aided" (CA), 35
Computer-aided design (CAD), 35, 188–191
Computer-generated images, 73, 74, 87
Computerization, 37, 66
Computers. *See also* Microcomputers; *specific topics*
  Jacques Ellul on, vii, 30, 31
  Bertrand Gille and, 30, 31
  Kenya Hara on, 190, 193
  Steve Jobs on, vii, 33, 88
  as "total technological object," 38, 39
  Sherry Turkle on, vii, 121
  virtual worlds and, 76, 77
Constructivism, 43, 97, 113
  aesthetic, 51
  defined, 43
  phenomenological, 47
Constructivist epistemologies, 43–44. *See also* Epistemological constructivism
Constructivist technology, 46
Contemporary technological system (CTS), 36
  Bertrand Gille and the, 29, 30, 34, 36
  nature of the, 32, 34–35
  question of the, 29–32
  Michel Volle on the, 34–36
Countess de Pange (alias de Broglie, Pauline), 61–63
Creative phenomenology, xix, 111
Creativity
  design and, xix, 185, 197, 198
  digital, 117, 118, 188, 191, 196
  industry and, 20, 22
  technology and, 20, 113, 182
Crime and design, 166
Critical eye vs. visionary eye, 5
Cultural value, technology as a, 17–23

Culture, 52
  technology and, 17
Cyberspace
  defined, 38
  digital ontophany and, 66
  rise of, 66
  uses of the word, 38, 79
Cynical design, 164

Dagognet, François, 18, 24
Darras, Bernard, 8, 109, 119–120
Death, 101
de Broglie, Pauline (alias Countess de Pange), 61–63
Debugging, 94, 95
Decoration, 155–157
Decorative arts, 153, 177
  defined, 159
  design and, 159
  England and, 159–161
  Germany and, 161
  ideals of the, 160
  industry, industrialization, and, 158–161
  modernity and, 158
  William Morris and, 158, 159, 177
  Hermann Muthesius and, 159–160
Dennett, Daniel, 137
Design, 19. *See also specific topics*
  as "a thing that thinks," 153, 183, 185
  concept of, 149, 150, 157
  deconstructing and rebuilding the word, 155–162
  definitions, 19, 165, 198
  essence/quiddity of, 174
  etymology of the word, 156
  goals/purposes, 178, 193, 198
  history of, 176, 224n15
  "It's so design," 155
  the lesson of (*see* Technology: as a cultural value)
  meanings and uses of the word, 155–157

Design (cont.)
  nature of, xix, 174, 177, 182, 198
  ontology of, 149
  "original sin" of, 170–172
  origin and birth of, 160, 161, 224n15
  origin of the word, 19, 156, 157, 159–162
  phenomenology of, 150, 151
  as a phenomeno-technological activity, 111–116
  philosophy of, xix, 149–150, 167, 174, 183
  political economy of, 166
  problems in the future of, 177–178
  terminology, 156, 162, 198–199
  theory of the effect of, xix (*see also* Design effect(s))
Design degree zero, 156
Designed object, 166, 197–198
  conditions under which an object becomes a, 174
  definition and nature of, 197, 198
  vs. other types of objects, 174
Designed subject, 166
Design effect(s), xix, 116, 173–178, 196–199. See also Digital design effects
  definitions and nature of, 174, 197, 199
  example of, 178
  types and dimensions of, 115–116, 150, 151, 174–178, 198
"Designer," 155, 180–181
Designers, 167. See also Industrial designers; *specific topics*
  as projectors, 115, 181
  twentieth-century, 185
Designing, defined, 167
Design perspective, 5
Design process, 90, 150, 168, 184, 196. See also *specific topics*
  computational matter and, 117
  defined, 197

fields of application of, 198
objects resulting from, 115
purpose, 198
Design projects, 4, 181, 182, 185
Design thinking, 150, 183–186. See also *specific topics*
Dialectics of the apparatus and of appearance, 53–60
Digital, the
  phenomenological trauma of, xviii, 65
  as a phenomenon, 10 (*see also* Digital phenomena)
  taking command, 32–39
  the unsustainable fragility of, 95
"Digital," use of the term, 80
Digital age, 12, 42, 104, 134, 145, 226n10
  dualism in the, 222n11 (*see also* Digital dualism)
  perceiving in the, 42, 144
  social connection in the, 98, 99
  Sylvie Leleu-Merviel on, 35
  video games and the, 92–93
  virtuality and, 93
Digital beings, phenomenality of, 42, 43, 111
Digital creative revolution, 191
Digital design
  defined, 117, 196
  vs. digitally aided design, 191
  intent of, 193
  interaction design and, 191
  origin of the concept of, 192
  toward, 187–196
Digital design effects, 117
  what they make possible, 117–120
Digital divide, generational, 63, 64
Digital dualism, 141–143, 222n11
  origin of, 144–146
  origin of the term, 143
Digital flows, 85, 100–101
Digital humanities, 4–5

Digitally aided design, 191, 195
   defined, 191
Digitally assisted design, 117
Digital monism, 145, 146
Digital natives, 63–64, 139
Digital noumena, 85, 91, 92, 94. *See also* Noumenon
   becoming phenomenal reality, 94
   phenomenalization of, 85
Digital ontophany, 48, 50–51, 66, 78, 80
   acculturation of perceptual structures to the new, xviii
   characteristics, 82–110
   creativity and, 111, 117
   digital design and, 117
   experiences and, 127–129
   vs. face-to-face ontophany, 121–122, 128–129
   limits, 129
   living in, 95
   nature of, 43, 50, 111
   shaping, 120
   telephonic ontophany and, 78, 98, 121–122, 127
   video game technology and, 119
Digital phenomena, 10. *See also under* Phenomenality
   can be annihilated, 103–105
   can be canceled, 101–103
   as copyable, 99–101
   as interaction, 89–93
   as otherphanic, 96–99
   as playable, 107–110
   as programmable, 86–89
   as simulation, 93–94
   as thaumaturgical, 105–107
   as unstable, 94–96
Digital relationships, 122
Digital revolution, xvii, 7–8, 23, 32, 125. *See also* Digital creative revolution; *specific topics*
   designing the, 120
   historical perspective on the, 27, 28, 34, 42, 65, 187
   nature of the, 5, 8, 34, 42, 85
   digital revolution as a digital revelation, 9–10, 126
   digital revolution as an ontophanic revolution, 8, 42, 65, 128, 145
   digital revolution as a phenomenological revolution, 125–126, 128
   digital revolution as a social revolution, 42–43
   digital revolution as philosophical event, 8, 42
   phenomenological violence of the, 81, 118
   philosophical understanding of the, 43
   and the philosophy of technology, 81–82, 118–119
   technological systems and the, 27
   tech revolutions and the, 27, 34, 58
Digital technological system, 36
Digital thaumaturgy, 106, 107
Digital virtuality, 71
Digitization (of thought), 37
"Disconnectionists," 121
"Double bind," 169
Drawing, 156–157, 173, 188–190
Dualism, metaphysical, 142, 145

Electronic revolution, 30
Éliade, Mircea, 48, 209–210nn21–22
Ellul, Jacques
   on computers, vii, 30, 31
   Bertrand Gille and, 14, 25, 30–32, 203n13
   Gilbert Hottois and, 16
   on the technological, 15
   technological revolution and, 32
   on technological/technical systems, 13–16, 21, 25, 31
Email, otherphany of, 134

Employment, 35
"Employment crisis," 36
Energy, 11
Enframing, 15, 21
Enthusiasm
  defined, 22
  the lesson of, 22
Entropy, 101
Eotechnical ontophany, 48, 49
Epistemological constructivism,
      44, 47. *See also* Constructivist
      epistemologies
Ethics and design, 167. *See also* Morality
      of design
Existence, 183. *See also* Being
  technology and, 56–58
Existential sphere, 112. *See also* Umwelt
Experience, defined, 47
Experience effect, 115, 174–175. *See also*
      Ontophanic effect
Experience-to-be-lived
  defined, 197
  design and, 174, 175, 198
  design effect and, 175, 197
  digital design and, 196
  experience effect/ontophanic effect
      and, 115, 174, 175
  projects and, 181
  use and, 197

Facebook, 108, 135, 142–143. *See also*
      Social media/social networks
Facebook "friends" and Facebook
      "friendship," 98, 122, 128,
      134–135, 146
Factitive intentionality, design and,
      114–119
Fetishism of technology, 16
Film. *See* Movies
Findeli, Alain, 115, 177–178, 181,
      223n4
Floridi, L., 95
Flusser, Vilém, 156, 167–168, 173

Form and function, aligning, 164
Forms, theory of, 177
Foster, Hal, 166, 173
Francastel, Pierre, 41
Frechin, Jean-Louis, 154, 192–193, 195
Free software movement, 21–22
Freud, Sigmund, 176, 179
Friendship, 135. *See also* "Weak ties"
Functional aesthetics, 161
Functionalism, 164
Functionalist aesthetics, 170
Function and form, 164

Game design, 118, 119, 196
  defined, 196
  interaction design and, 119
  web design and, 119
Gameplay, 109, 110, 119. *See also* Play;
      Video games
Gameplay design, 110, 119
Games, 109, 110. *See also* Video games
Gamification, 108
Generational digital divide, 63, 64
Genvo, Sébastien, 108
Geometry, 41
  non-Euclidean, 41, 42, 208n2
Giedion, Siegfried, 37, 87
Gille, Bertrand, 33
  and changes in technological systems,
      28
  compared with Thomas Kuhn, 28
  computers and, 30, 31
  on electronic revolution, 30
  Jacques Ellul and, 14, 25, 30–32,
      203n13
  employment crisis and, 36
  The History of Techniques, 11, 14, 30
  on information and communication
      technologies techniques, 33
  on innovation, 38
  on mechanization, 27
  ontophanic milieux and, 138
  and philosophy of technology, 13–14

technical combination and, 24
technical concatenation and, 12
technical systematicity and, 14, 25
technological combination and, 11, 12, 24, 34–35
on technological complex, 12–13
technological revolution and, 29, 32
technological systems and, 11–14, 28, 31, 138
  the contemporary technological system (CTS), 29, 30, 34, 36
  terminology and, 36, 203n13
  Michel Volle and, 34, 36
  Jacob von Uexküll and, 138
Global village, 66
Good Design, 170
Granger, Gilles-Gaston
  actuality, nonactuality, and, 67, 84
  on mathematics, 84
  noumena and, 84
  on regimes of reality, 83
  science and, 83, 84
  on virtual phenomena, 83, 84
Graphical user interface (GUI), 71–73, 195
GRiD Compass (first laptop), 89–90, 191
Gropius, Walter, 160
Guillaud, Hubert, 36

Habermas, Jürgen, 15
Halo effect, 119–120
Hara, Kenya, 176
  on computers, 190, 193
  on design, 153, 174, 181, 224n15
  emptiness and, 176, 192
Harvard Mark II, 94
Hegel, Georg Wilhelm Friedrich, 109
Heidegger, Martin, 9, 15, 21, 52
Heilbrunn, Benoît, 165
Heraclitus, 101
Hierophany, 209–210nn21–22
Hopper, Grace, 94
Hottois, Gilbert, 16, 18

Household appliances, electric, 30
Human-computer interaction (HCI), 90, 95
Humanists, 126
Humanities, 4–5, 193
Human-machine interfaces (HMI), 194
"Human needs is the place to start" (step in design thinking), 184
Huyghe, Pierre-Damien
  aesthetics and, 56–59, 211n43, 211n47
  on apparatus, 56, 57
  on phenomeno-technology, 58–59
  on technology, 56–59, 211n47
Hyper-industrial era, 19–20

Ideation time, 184
iMac, 175
Images
  screens as worlds of, 66
  types of, 68
Image synthesizing, 69, 74, 87, 142
Imaginary, the, 70, 76, 79, 81. *See also* Metaphysical imaginary
  defined, 70
  and the virtual, 70, 76, 79–81, 83, 142
Imaginary metaphysics, 76. *See also* Metaphysical imaginary
Imagination, 70
Immersion, 120–123
  in digital interfaces, 50–51, 66
Immersive environments, 74, 77, 78
Implementation (step in design thinking), 184
Industrial aesthetics, 161, 162, 176
  defined, 161
  France and, 19, 161, 162, 176
  origin of the term, 161
Industrial design, 167, 169, 170, 172, 190, 192, 199. *See also* Industrial aesthetics
  digital design and, 196

Industrial design (cont.)
  ethics of, 167
  harmfulness of the profession, 167
  marketing and, 163–164
  nature of, 19
  Victor Papanek and, 167, 170
  terminology, 19, 162, 198–199
Industrial designers, 90, 160, 163, 167, 169, 170, 172, 191
  art and, 157
  Victor Papanek and, 167
Industrialization, 37, 163, 166. See also Industrial Revolution
  computerization and, 37
  decorative arts and, 158, 160
  design and, 157, 163, 166, 169, 199
  technology and, 23
Industrial production, 15, 157, 161, 168, 169, 173, 175, 199
Industrial Revolution. See also Second Industrial Revolution/Technological Revolution
  first, 13, 27, 28, 32, 34, 37
  third, 32
Industry
  arts applied to, 159 (see also Applied arts)
  as first field of applied design, 199
  first great collaboration between art and, 160
Informatics, 31, 32, 37, 72
Information and communication technologies (ICT) sector, 193
Information revolution, 32
Information technology (IT) sector, 21–22, 193
Innovation, 23
"Innovation without jobs," 36
In-person otherphany, 132
Instruments, 45
Intelligence, 44
Interaction design, 191–195. See also User interface design
  defined, 192, 194
  origin of the term, 192, 194
Interface design. See User interface design
Interfaced perception, 117
Interfaces, 120. See also Graphical user interface; User interface
  phenomeno-technology and, 79, 80, 85
"Interface," use of the word, 90
Intermediate worlds/intermediate realities, 75
Internet, 3–4, 7, 10, 38–39
Invention, 23
iPad, 123, 195, 196
iPod, 100, 101
IRL (in real life) fetishism, xviii

Jobs, Steve, 20, 27, 88, 89, 91
  Apple products and, 20, 21
  on computers, vii, 33, 88
  on Macintosh's success, 88
Jones, Sheilla, 46
Jouin, Patrick, 178, 180–181, 191, 199
Joyn table, 116
Judgments, 144
Jurgenson, Nathan J., 135, 142–143

Kant, Immanuel, 14, 43
  noumena and, 45, 83
  phenomena and, 43, 45, 82–83
  reason and, 16, 43
Kitchens, designer, 155, 166
Kuhn, Thomas S., 28

Lalande, André, 67
Latour, Bruno, 112, 125, 126
Lavoisier, Antoine-Laurent, 104
Learning by doing (step in design thinking), 184
Lebovici, Serge, 97
Le Corbusier, 176

Index

Leleu-Merviel, Sylvie, 35, 88
Le Moigne, Jean-Louis, 43
Leroux, Yann, 10, 106
Lessig, Lawrence, 22
Lévy, Pierre, 38
  on cyberculture, 202n5
  on programming, 88
  on technology, 48, 209n20
  on transcendental history, 52
  on the virtual, 76, 142
Life and death, 101
"Life on screen," 7, 91
Local area networks, 38
Loewy, Raymond, 19, 163, 164
  design and, 19, 164, 165, 170
  on industrial world design, 161
Loos, Adolf, 166, 176, 177
Loving otherphany, 132–133
Ludification, 108, 109

Maeda, John, 143, 164, 179, 187, 193
Make-being and make-making, 115, 116, 118
Marcuse, Herbert, 15
Marketing, 163–166
  defined, 164
Marketing design, 164, 165, 170, 171
  defined, 171
Marx, Karl, 27, 157–158
Materials, 11
Mathematics, 75, 88, 89
  Gaston Bachelard and, 45, 46
  Gilles-Gaston Granger and, 84
  noumenality and, 45, 46, 83, 85, 86, 92
  quantum physics and, 41, 45, 46, 83
  Philippe Quéau and, 75, 87
  and the virtual, 84, 87
Mathias, Paul, 38–39, 85, 87
Matrix, The (film), 33, 145
Measuring instruments, 45
Mechanization, 27–29, 33–37
Mechanized ontophany, 48–50

Metaphysical aura. *See* "Aura"
Metaphysical imaginary, 66, 79–80.
  See also Imaginary; Imaginary metaphysics; Metaphysics: of the image
Metaphysicians, 81–83
Metaphysics, 9, 70, 75, 141–142, 145
  of the image, 76–78 (*see also* Metaphysical imaginary)
  new (*see under* Virtual)
Microcomputers, 12, 24, 30, 34, 37, 66, 72, 73, 89
Midal, Alexandra, 159, 163, 171
Milieu, 138, 139
Miller, Paul, 123, 146
Misology, 16–17
Misotechnology, 16–17
Moggridge, Bill, 89–90, 185–186, 191–192, 194
Morality of design, 171–172. *See also* Ethics and design
Morris, William, 158–159, 165, 167, 177
Movies, 92, 93
Muthesius, Hermann, 159–161

Nature, 51
Networks, 31–34, 37–39, 85–86, 96. *See also* Social media/social networks
Nietzsche, Friedrich, 22, 76
Norman, Donald ("Don"), 185–186
Noumenality
  mathematics and, 45, 46, 83, 85, 86, 92
  pure, 83
Noumenal perspectives, 46
Noumenal world, 45
Noumenon
  Gaston Bachelard on, 45, 46, 82, 83
  definition and nature of, 45, 83–85
  the digital phenomenon as a, 82–86, 91 (*see also* Digital noumena)
  vs. phenomenon, 45, 91
  and the virtual, 84, 94

"Noumenon and Microphysics" (Bachelard), 44, 53
Nuclear physics, 44–45

Online friendship, 133–135
Ontology, 8–9, 42, 52, 96, 153
Ontophanic conditions, 52, 121, 138
Ontophanic devices. *See* Phenomeno-technological devices
Ontophanic effect, 115, 150, 175. *See also* Design effect(s): types and dimensions of
Ontophanic feeling, 139
Ontophanic future, interactive situations and our, 120–123
Ontophanic grid, 52, 55
Ontophanic matrix(ces), xviii, 56, 64
  defined, 8
  digital interfaces as a new, 79
  digital medium as the new, 134
  nature of, 8, 51
  phenomenological aura given off by, 127
  social networks as, 99
  technology as, 47–53, 58, 63
Ontophanic milieux, 138, 139
Ontophanic revolution(s), 56, 60, 63, 145
  defined, 145
  digital revolution as an, 8, 42, 65, 128, 145
  technological revolutions and, 52
Ontophany, 10, 81. See also specific topics
  being-in-the-world and, 50–51, 63, 112, 122, 138
  Mircea Éliade on, 48, 209n21
  eotechnical/premechanical, 48–50
  etymology of the word, 145
  the fabrication of, 112–114
  face-to-face, 121–122, 128–129
  meanings and uses of the word, 48, 209n21
  of music, iPod as generating a new, 100
  phenomenality and, 43, 48, 51, 55, 61, 65, 66, 79, 97, 100–102, 111, 112, 123, 131–133, 136
  as a phenomeno-technological result, 112
  phenomeno-technology and, 51, 55, 59, 61, 65, 94, 98, 99, 112
  technical, 48
  technological, 43, 48
  theory of, 210n24
Operations, 11
Ornament, 176, 177
"Ornament and Crime" (Loos), 166
Other, the. *See also* Otherness
  as a technologically produced phenomenon, 131–133
Otherness, 131
  experience of, 131–133
  ontophany and, 98, 131–133, 135–136
  phenomenalization of, 133, 136
  transcendental technicity of, 131–133, 135
  types of, 131
Otherphanic feeling, 137–139
Otherphanic phenomena, digital phenomena as, 96–99
Otherphany, 98, 131–133
  friendship and, 134–135
  in-person, 132
  loving, 132–133.
  nature of, 98, 131
  otherness and, 98, 131–133, 135–136
  telephonic, 132
  toward digital, 133–135

Papanek, Victor, 167, 170
Paradigm shifts, 28, 163
Paradoxical injunction, 169
Perception, 144
  in the digital age, 42

Perceptual culture, 41
Perceptual environment, 56, 117, 138.
    See also Umwelt
Perceptual surrounding, 52
Perrault, Dominique, 113
Perspective, laws of, 51
Phenomena, active development of, 45
Phenomenality, 46–48, 52, 55, 97, 101, 106
    ability of the apparatus to produce, 57
    culture generating, 52
    of design, 174
    digital, 66, 79
    digitally centered, 123
    natural, 54
    of nature, 51
    ontophany and, 43, 48, 51, 55, 61, 65, 66, 79, 97, 100–102, 111, 112, 123, 131–133, 136
    otherness and, 131–133, 135–136
    of our relationship to others, 60, 61 (see also under Otherness)
    of phenomena, 51, 112
    technical, 59
    technological, 54
    of the unstable, 96
    of the world, 60, 65
Phenomenal phenomenality, 48
Phenomenological archaeology, 144
    defined, 144
    of technologies, xvii–xviii (see also Digital dualism: origin of)
Phenomenological aura, 122, 127–129.
    See also "Aura"
Phenomenological judgment, 144
Phenomenological revolutions, 41
Phenomenological trauma of the digital, xviii, 65
Phenomenology
    phenomeno-technology and, 44, 45
    of technologies, historical, xvii–xviii

Phenomenon, defined, 45
Phenomeno-technological activities, 115–116
Phenomeno-technological apparatuses, 55, 59, 65. See also Phenomeno-technological devices
Phenomeno-technological casting, 52, 66, 98, 126, 127
Phenomeno-technological construction, experience as a, 51, 113, 118
Phenomeno-technological constructivism, 97
Phenomeno-technological devices, 59, 100, 112. See also Phenomeno-technological apparatuses
Phenomeno-technological intermediaries, 85, 91, 98
Phenomeno-technological mediation, 133, 136
Phenomeno-technology, 43–48, 55–56, 63, 64, 107, 132–133
    of art, 55, 58
    Gaston Bachelard and, 14, 44–46, 53, 55, 58
    Walter Benjamin and, 53, 55, 56, 58, 59
    concept of, 44, 53
    constructivism and, 44, 46
    general, 55, 57–59, 65
    history of, 119
    interfaces and, 79, 80, 85
    meaning of, 44, 46, 56, 65
    nature of, 44, 56
    noumenology and, 46
    of objects, 115
    ontophany and, 51, 55, 59, 61, 65, 94, 98, 99, 112
    origin of the concept, 44, 53
    phenomenal reality and, 46
    psychoanalytic treatment and, 128
    science and, 45, 46, 55
    virtuality and, 80, 93–94

Philosophy, 8, 9, 17. *See also specific topics*
Photographic images, 68
Photography
 aesthetics/beauty and, 54–55
 art and, 53–54
 and the aura, 54, 56, 129
 Walter Benjamin and, 53–57, 129
Physics
 modern, 45, 46 (*see also* Quantum physics)
 nuclear, 44–45
Piaget, Jean, 43–44
Plato, 145
 on intelligible forms, 75, 76
 intermediate realities, 75
 Philippe Quéau and, 75, 76
 and the virtual, 75–78, 82, 145
Platonic metaphysics, 75–78, 145
Plato's cave, 82, 145
Play, 107–110, 117, 119
Possible, the, 83, 84. *See also under* Digital design effects
Possible experience, 83, 111, 112, 127, 133, 134, 136. *See also under* Digital design effects
Possible experience-of-the-world, 49, 50, 100, 112–114, 122–123
Postmodern design, 19
Postmodern individuals, 166
Postmodern societies, 168, 176
Potential/potentiality (*dunamis*), 66–68, 142
Premechanical/eotechnical ontophany, 48–50
Premechanical revolution, 27, 28
Probable, the, 83, 84
Production of effects, 150
Programming languages, 86
Projecting in design, 181
Projectors, designers as, 115, 181
Projects, design, 4, 181, 182, 185
Psychic acceleration, 107

Psychoanalysis, 144, 176, 179
 French, 70
 and the virtual, 70, 71
Psychoanalytic treatment, 128–129

Qualia, 137
Quantum beings, 85, 127
Quantum noumena, 44, 45, 82–85
Quantum perception, 127–128
Quantum physics, 41, 42, 46, 82–84. *See also* Quantum noumena
 Gaston Bachelard on, 44–46, 82–84
 mathematics and, 41, 45, 46, 83
Quarante, Danielle, 224n15
Quéau, Philippe, 73–75
 image synthesizing and, 74, 87, 142
 Plato and, 75, 76
 on the symbolic, 75
 on virtual images and the virtual, 65, 67, 74, 75, 87, 106, 142
 writings, 73–74

"Real," 142. *See also under* Virtual
Real images, 68
Reality
 concept of, 84
 three regimes of, 83 (*see also* Possible; Virtual)
Reason
 hatred of (technological), 16–17
 Kant on, 43
Reduplication, 174
Remaury, Bruno, 156
Renaissance, 13, 27, 28, 41–42, 49, 51, 88
Reverie
 Gaston Bachelard on, 77, 78
 end of the, 77–80
 of the unreal, 81, 94
 of the virtual, 77–80, 83, 94, 97–99
Reversibility, ontophany of, 102
Rey, P. J., 135
Rietveld, Gerrit, 179

Rifkin, Jeremy, 32
Rousseau, Jean-Jacques, 189
Ruskin, John, 158
Rutherford, Ernest, 44

Science
　as phenomeno-technological, 45, 46
　philosophy of, 18
Science-fiction films, 33. *See also Matrix, The* (film)
Scientific revolutions, 28
Screens, fear of, xviii
Searles, Harold F., 169
Second Industrial Revolution/Technological Revolution, 28, 29, 32, 37
Sector, 12–13. *See also* Information technology (IT) sector
Security patches, 95
"Seeing things at (inter)face value," 77
Self-centered world, 52, 112. *See also Umwelt*
Séris, Jean-Pierre, 18
　on beauty, 176
　on misotechnology, 16–17
　on postmodern design, 19
　on technology, 11, 16, 18–21, 23, 24, 105, 176
Serres, Michel, 111
Short Treatise on Design, A (Vial), xviii–xix, 149, 154
Simon, Herbert, 181
Simondon, Gilbert, 15, 18, 57, 125
　on machines, 8, 17
　on milieu, 138, 139
　on philosophy of technology, 14
Simulation, 73–75
　computational, 69, 73, 74
　digital phenomena as, 93–94
　meanings and uses of the word, 73, 188
　Philippe Quéau on, 74
　virtual machines and, 69–71

Simulation and Its Discontents (Turkle), 188–189
Simulation culture, 78, 86
　Sherry Turkle on, 73, 78–79, 188
Sloterdijk, Peter, 9, 52, 112, 116, 126
Social imaginary, 144
Socialism, 158, 169. *See also* Marx, Karl
Social media/social networks, 50, 96–97, 128, 133–134, 142, 143. *See also* Facebook
Social reform effect, 177. *See also* Socioplastic effect
Sociodesign, 177
Socioplastic approach, 181
Socioplastic effect, 150, 177, 178. *See also* Design effect(s): types and dimensions of
　defined, 177
Socioplastic forms, 177
Socioplastic (societal) effect, xix, 150, 177, 178, 185
Sottsass, Ettore, 153, 170–172
Souriau, Étienne, 159, 162
Sphere, 112
Stallman, Richard M., 20–22, 193
Starck, Philippe, 164, 173, 176, 179
Stiegler, Bernard, 19, 139, 164–165
Stoicism, 103
Stylique, 162
Synthesized images. *See* Image synthesizing
System, concept of a, 14, 16, 25. *See also* Technological/technical systems

Tech-born, 64
Tech natives, 139
Technical concatenation, 12–13, 51
Technical ensemble level, 12
Technical ensembles, 12–14
Technical instruments, 45
Technical structure level (technical combination), 12

Technical systems. *See* Technological/technical systems
Technical/technological combination(s), 9, 11–13, 24
  Bertrand Gille and, 11, 12, 24, 34–35
  as horizontal, 25
  levels of, 11–13, 25
  Michel Volle on, 34–35, 39
Techniques, 11
Techno-fetishism, 16
Technological, the, 15, 25, 42, 51. *See also specific topics*
  Jacques Ellul on, 14, 15 (*see also under* Ellul, Jacques)
  as ontophanic matrix, 48, 51 (*see also under* Technology)
Technological convergence, 23–25, 32
Technological milieu, 138, 139
Technological mind, new, 7–8
Technological revolutions. *See also* Second Industrial Revolution/Technological Revolution
  as materials revolutions, 85
Technological shift, 58
Technological/technical systems, 11, 27
  in the age of "technology," 23–25
  Jacques Ellul on, 13–16, 21, 25, 31
  Bertrand Gille and, 14, 25, 27-34-36, 138
  history of, 27–29
  nature of, 11–14, 17
  new, 32–39
Technologies, historical phenomenology of, xvii–xviii
Technology, 11, 41. *See also specific topics*
  being-in-the-world and, 52–53, 63, 125, 138
  creativity and, 20
  as a cultural value, 17–23
  essence of, 59
  generates phenomenality, 46 (*see also* Phenomenality)
  history of
    digital revolution and the, 27
    lessons of the, 25, 32–33, 98
    tech history of the West, 28
  as ontophanic matrix, 47–53, 58, 63
  philosophy of, 8, 15–17, 21, 25, 46–47, 56
  Apple products and, 20
  digital revolution and, 81–82, 118–119
  Bertrand Gille and, 13–14
  history of technologies and, 14
  Pierre-Damien Huyghe and, 59
  techno-transcendental phenomenology and, 56, 59
  writings on, xvii
  and the question of being, 8–10
  as a structure of perception, 43
  terminology, 23, 203n13
  use of the word, 23, 24
Technology revolutions. *See* Digital revolution; "Tech revolutions"
Technoperceptual milieux, 139
Technophany, 210n22
Technophobia, 16–18, 20, 22, 145
Technoscience, 16, 23
Technosystems and technosystematicity, 14, 25
Techno-transcendentality, 131
Techno-transcendental phenomenology, 53, 56, 59
Techno-transcendental structures, 43, 48, 50, 52, 61, 66, 99, 112, 113, 115, 137
"Tech revolutions," 27, 28. *See also* Digital revolution
Telephonic ontophany, 78, 98, 127, 132
  digital ontophany and, 78, 98, 121–122, 127
  model of, 60–64
  as a naturalized perceptual culture, 63
  telepresence and, 98, 122
Telephonic otherphany, 132

Index

"Telephonization of life," 62
Textile industry, 12
Thackara, John, 103
Tisseron, Serge, 63, 70–71, 83
Total social fact, 18, 38, 187
Total technological object, 31, 38, 39
Transcendence, 23, 24. *See also* Techno-transcendental structures
 "black," 16
Transcendental history, 52
Transcendental structures, 43
Transcendental technicity, 133
Transistors, 12
Triclot, Mathieu, on video games, 93, 101–102, 118–119
Turkle, Sherry, 70, 78–79, 188–189
 on CAD, 188, 190
 on computers, vii, 121
 on immersion, 120–121
 on simulation, 188–189
 on simulation culture, 73, 78–79, 188
Twitter, 143. *See also* Social media/social networks
Twitterfly effect, 143

Uexküll, Jacob von, 138, 139. *See also* Umwelt
Umwelt, 52, 56, 65, 112, 117, 138, 139
Use (of designed object), defined, 197
User-centered design, 192, 194. *See also* Interaction design
User interface, 195
User interface design, 90, 191, 192, 194

Value judgments, 144
Van de Velde, Henry, 160, 176
Varenne, Franck, 193
Vélib', 178
Verplank, Bill, 90, 192
Video games, 70, 92–93, 107–108, 118, 119, 123. *See also* Games
 design effect and, 196
 game design and, 118, 119, 196

vs. movies, 93, 101
 Mathieu Triclot on, 93, 101–102, 118–119
Viénot, Jacques, 161
Vignola, Robert, 60, 210n28
Virtual, 10, 66, 69
 Aristotle and the, 67–68
 from a culture of programming to a culture of the, 73
 definitions and meanings of the word, 67–71, 73, 74, 81, 141, 213n14
 genealogy of the, 66
 philosophy, optics, computing, and psychoanalysis, 66–71
 misunderstanding of the, 71
 from the neometaphysics to the vulgate of the real and the, 72–78, 94
 origin and etymology of the word, 66–67
 vs. real, 141–142
 reverie of the, 77–81, 83, 94, 97–99
 uses of the word, 80, 81, 142
Virtual images, 69, 73, 79, 82, 145
 definition and nature of, 68, 69, 73–76, 79, 87
 Plato and, 75, 76, 82 (*see also under* Plato)
 Philippe Quéau and, 74, 75, 87, 106
Virtuality, 77, 80, 82–84, 93–94
 digital, 71
 dimension of, 71, 82
 phenomeno-technology and, 80, 93–94
 psychic, 70–71
 terminology, 83
Virtualization, 142
Virtual machine, 69
Virtual memory, 69, 93
Virtual reality, 10, 69, 78
Virtual worlds, 76, 80, 85
 computers and, 76, 77
 "cyberspace" and, 79

Virtual worlds (cont.)
  defined, 74
  nature of, 87
  Philippe Quéau on, 74–76
  in video games, 93
Virtus, 67, 142
Visionary eye vs. critical eye, 5
Volle, Michel, 33–39
  on contemporary technological
    system (CTS), 34–36
  on technological combination,
    34–35, 39
von Hagens, Gunther, 180

"Weak ties," 122, 128. *See also*
    Friendship
Web design, xix, 90, 119–120, 192, 196
  defined, 196
Werkbund, 160, 161, 172
Whatever Works (film), 101
Wikipedia, 4
Wilde, Oscar, 51
Workstation model, 72, 120
Wozniak, Steve, 88